Case Studies in Pollution Control Measures in the Textile Dyeing and Finishing Industries

Other Titles of Interest

ATKINS AND LOWE:

Economics of Pollution Control in the Non-Ferrous Metals Industry

Pollution Control Costs in Industry

Case Studies in Pollution Control Measures in the Textile Dyeing and Finishing Industries

A handbook in non-technical language of essential information on the economics of control, the problems and their solutions

by

M. H. ATKINS

Department of Business and Professional Studies
Trent Polytechnic, Nottingham

and

J. F. LOWE

School of Management
University of Bath

PERGAMON PRESS

OXFORD · NEW YORK · TORONTO · SYDNEY · PARIS · FRANKFURT

U.K.	Pergamon Press Ltd., Headington Hill Hall, Oxford OX3 0BW, England
U.S.A.	Pergamon Press Inc., Maxwell House, Fairview Park, Elmsford, New York 10523, U.S.A.
CANADA	Pergamon of Canada, Suite 104, 150 Consumers Road, Willowdale, Ontario M2J 1P9, Canada
AUSTRALIA	Pergamon Press (Aust.) Pty. Ltd., P.O. Box 544, Potts Point, N.S.W. 2011, Australia
FRANCE	Pergamon Press SARL, 24 rue des Ecoles, 75240 Paris, Cedex 05, France
FEDERAL REPUBLIC OF GERMANY	Pergamon Press GmbH, 6242 Kronberg-Taunus, Pferdstrasse 1, Federal Republic of Germany

First edition 1979

British Library Cataloguing in Publication Data

Atkins, M H
Case studies in pollution control measures in the textile dyeing and finishing industries.
1. Textile industry — Environmental aspects
2. Pollution
I. Title II. Lowe, J F
628.5 TD195 78-41074

ISBN 0-08-022457-1

In order to make this volume available as economically and as rapidly as possible the author's typescript has been reproduced in its original form. This method unfortunately has its typographical limitations but it is hoped that they in no way distract the reader.

Printed and bound at William Clowes & Sons Limited
Beccles and London

To Gillian, Margaret, Sally and all women who are still ladies

Contents

List of Tables, Figures and Cases

Preface

This book arises out of research conducted by ourselves under a grant from the Department of the Environment into pollution control costs in the Textile Dyeing and Finishing Industry and we are grateful for their permission to publish some of the material that was contained in our report on that study. The conclusions and views published here are,however, those of the authors alone and do not necessarily represent those of the Department of the Environment.

In the course of our research we visited a number of firms and we wish to record our thanks to them for their co-operation in the original research. Special thanks are due to those who have given their permission for the publication of the anonymous case studies in this book.

We would also wish to express our special thanks to Dr. R.Richardson of Courtaulds Ltd. who has read and commented upon the case studies and also to Dr. T.H. Morton who has commented upon the technical material. Both have provided us with a wealth of practical experience that we, as economists, were not able to draw upon. Thanks must also go to Miss Ros Keen who worked with us on the original study and who helped write our original research report. All errors, omissions and views expressed remain, however, the responsibility of the authors alone.

Preface

[illegible]

Introduction

Pollution and pollution control are part of the "facts of life" of the modern business enterprise; there are probably few capital expenditure decisions taken that do not have to take pollution factors into consideration.

The way in which firms respond to pollution controls has been an area very much under-researched by economists at least. We have attempted to redress this balance - our first book* was a study of pollution control costs in largely small and medium-sized firms mainly in the East Midlands region. This study and a related one [+] refer to national samples and include large firms.

This type of research is not easy. A technical knowledge of the processes involved is required and one soon learns that whilst superficially firms may seem similar, specific differences in their processes, products, age or location may contribute to making their pollution problems and control costs quite different. In presenting these case studies, a primary purpose is to emphasise the heterogeneity of the industry and not its homogeneity. Even if *firms* were similar, pollution controls vary spatially so that, for instance, requirements on colour of effluent discharge may vary in different regions and thus impose specific problems on some firms. These twenty or so cases therefore span almost the entire spectrum of firm size, organisational structure and product and process type. The reader may discern few common threads but then a key message is the basic heterogeneity a single industry can display.

Many of the firms in these case studies have experienced very large increases in water and effluent charges in recent years. The historical background to these developments lies in the 1973 Water Act which led to the formation of the Water Authorities; Section 30(5) of this Act requires Water Authorities to take such steps as

---

* Pollution Control Costs in Industry - an Economic Study, Pergamon Press 1977

\+ The Economics of Pollution Control in Non-Ferrous Metallurgy, Pergamon Press 1978

will ensure that, as from a date not later than 1 April 1981, their charges are such as not to show undue preference to, or discriminate against, any class of persons. Historically parts of the textile industry have been treated favourably in order to bring employment to particular localities. Now Water Authorities are moving towards charging firms the total overall costs of disposing of their effluent. In the past charges have been based on Biological Oxygen Demand (B.O.D.) but the move towards Chemical Oxygen Demand (C.O.D.) based charging (which was jointly recommended by the National Water Council and the CBI), may also have accentuated charging increases in some cases. It is also possible that this move (charging based on C.O.D. rather than B.O.D.) may improve the reliability of assessments - the analytical techniques for assessment of B.O.D. may have been unreliable in some cases. Further, of course, since the move in 1973 towards charging the full cost of effluent disposal, the rises in charges have been influenced by the general level of inflation since then. In the early stages of their formation, water authorities may not have been able to indicate future possible trends in their costs and consequent charges.

The disposal of trade effluent is subject to consents imposed by the Water Authorities and in some cases these consents have changed. A Water Authority can review consents every two years. It may decide to relax standards if a case can be made for this. This might be done on the basis of changes in the work-load at the firm and changes in the capacity and overall load at the sewage works. There is also now a move, in preparation for the implementation of the relevant part of the Control of Pollution Act, to bring consents into line with discharges along the lines of the National Water Council's Consultation Paper, "Review of Discharge consent conditions" of 9/2/77. Some earlier agreements taken over by the Water Authorities were only for a limited number of years but in those cases where Water Authorities took over legal agreements, these had to be adhered to.

Chapter I

The Organisation and Structure of the Industry *

INTRODUCTION

The textile finishing industry is distinguished from other parts of the textile industry by the processes it involves and is defined in this study as the bleaching, dyeing, printing and finishing of all types of textiles. However in commercial terms finishing is not often regarded as a separate industry and textile activities tend to be split into industrial groups by the type of raw materials used. Traditionally three main divisions have been distinguished; the woollen and worsted section, cotton manufacturing and the knitted goods and hosiery trade. The introduction of man-made fibres and the formation of groups with interests in all sectors of the textile industry has tended to blur these historical boundaries.

This chapter examines the structural characteristics of the textile finishing industry and recent changes in market structure, to provide an economic context within which the case studies can be considered. The situation in Britain is discussed in detail first and some comparisons are then made with the United States and E.E.C. countries.

THE TEXTILE FINISHING INDUSTRY IN BRITAIN

The relative importance of the textile industry has declined through most of the twentieth century. The decline of the cotton industry was due largely to increased self-sufficiency and competition in former export markets as well as the growth of imports from low cost producers abroad. The woollen and worsted sectors declined less in the first few decades of this century, partly because there was less reliance on exports to developing countries and on easily manufactured products, but have reduced in importance since the 1950's. The knitted sector grew rapidly in the 1963-68 period and man-made fibres have taken an increasing share of the market since their introduction. As most textile products have to be finished, activity in the textile finishing industry tends to reflect these general trends.

---

* This chapter was written by E.J. Morgan of the University of Bath.

TABLE 1.1 NET OUTPUT AND EMPLOYMENT IN THE TEXTILE INDUSTRY

	Net Output (£ million)		Employment (000's)	
	1963	1974	1963	1974
All manufacturing	10,820	32,391	7,952	7,680
All textiles	792.4	1837.8	749.3	576.1
Textile finishing	65.9	121.4	62.9	44.6

Sources: Census of Production and Department of Employment Gazette.

Table 1.1 shows the relative decline of the textile industry recently in the context of all manufacturing. In 1963 all textile activities together accounted for 7.3 percent of all manufacturing net output and 9.4 percent of employment, but by 1974 their share of output had fallen to 5.7 percent and 7.5 percent respectively. The share of cotton goods in total textile output fell by 2.9 percent in the period 1963 to 1974 and the share of woollens fell by 6.8 percent. Although net output in textile finishing approximately doubled between 1963 and 1974, its share of total textile output fell from 8.3 percent to 6.6 percent, and employment showed absolute as well as relative decline.

The Census of Production shows that the most important activity among establishments with twenty five or more employees primarily engaged in textile finishing is the processing of cotton and man-made fibres which accounted for 55 percent of sales in 1973.* The knitted goods sector was second in importance with 29 percent of sales and the finishing of wool and worsted products (excluding finishing at weaving mills) accounted for 11 percent of activity.

The changing pattern of finishing production in the cotton and man-made fibres sector is given in Table 1.2. Piece bleaching and piece dyeing declined by about one third between 1965 and 1976, while machine printing more than halved. There were, however, important exceptions to the general trend. There has been a consistent upward movement in screen printing and the output of bleached and dyed mixture fibres has increased, reflecting the recent popularity of synthetic and cotton blends. Other processes such as easy care finishing, permanent pleating and proofing have also gained in importance.

---

* The Census treats textile finishing as a separate industry (MLH 423) including all establishments wholly or mainly engaged in the finishing of yarns and woven and knitted fabrics. The finishing of fabrics at weaving establishments is not included. Note that finishing at establishments when finishing is not the main activity is also excluded, a qualification that is particularly important when studying textile finishing using Census data, as the largest firms are vertically integrated with multiprocess plants. (1)

TABLE 1.2 OUTPUT IN COTTON AND ALLIED FIBRE FINISHING

(million linear metres)

	1965	1976
Piece bleaching		
Cotton	396.4	237.7
Other	45.6	58.9
Piece dyeing		
Cotton	260.5	118.8
Man-made	296.4	199.3
Mixtures and other	48.5	82.1
Machine printing		
Cotton	176.5	68.7
Man-made	36.5	11.5
Mixtures and other	19.8	19.3
Screen printing	31.2	122.0
Other processes	42.1	66.5

Most finishing in Northern Ireland is excluded.

Source: Textile Council Quarterly Statistical Review.

Regionalisation

The textile industry is heavily concentrated in three main regions. The spinning and weaving of cotton and allied textiles are mainly carried out in Lancashire and the North West, the woollen industry is centred in Yorkshire and the knitwear and hosiery trades are found mainly in the Midlands. Textile finishing activity is less regionally concentrated; as almost all textile products have to be finished at some stage, the pattern of employment is nearer to that of the textile industry as a whole than any particular sector. In 1975 the North West and East Midlands each had 28 percent of textile finishing employment and 18 percent was in Yorkshire and Humberside (2).

Concentration

One of the most important structural characteristics of an industry is the level of concentration as this provides some indication of the extent of competition within the industry. Table 1.3 allows comparisons between the recent development of concentration (measured by the share of the largest five firms ranked by employment in total employment) in textile finishing and other sectors of the textile industry. Concentration in textile finishing appears to have been fairly low, with the largest six firms accounting for only 36 percent of employment in 1972. However this finding must be interpreted carefully in view of the dominant position of the large vertically integrated concerns in many types of textile finishing.

TABLE 1.3 FIVE FIRM EMPLOYMENT CONCENTRATION RATIOS IN SELECTED TEXTILE SECTORS

Industry	1963	1968	1972
Production of man-made fibres	99	100	97[a]
Spinning and doubling on the cotton and flax systems	32	43	51[a]
Weaving of cotton, linen and man-made fibres	14	29	37
Woollen and worsted	18	21	26
Hosiery and other knitted goods	18	30	38
Textile finishing	28	37	36[a]

a Relates to the largest six enterprises.

Sources: Census of Production
1963 estimates by R. Clarke, National Institute of Economic and Social Research.

By far the most concentrated sector is the production of man-made fibres where a near duopoly situation has existed for some time, with Courtaulds and ICI as the leading producers. In 1972 Courtaulds had a monopoly of domestic rayon production, and was responsible for 80 percent of acetate fibres, 60 percent of acrylic fibres, 20 percent of nylons and 5 percent of polyester output in Britain. ICI dominated polyester and nylon production with shares of 80 percent and 60 percent respectively. (3)

There was a substantial increase in concentration in the cotton and knitted sectors between 1963 and 1968, largely due to the intervention of Courtaulds and ICI in these two areas, particularly the cotton industry, to secure the future of their markets for man-made fibres. However, despite the increasing dominance of the major firms, concentration in 1968 in these subsectors of the textile industry still remained less than in manufacturing industry as a whole. The major firms in the cotton industry are Courtaulds, Carrington Viyella (in which ICI owns 64 percent of the equity) and Tootal (in which both ICI and Courtaulds have 8 percent stakes). In hosiery and knitwear Courtaulds is the leading firm producing warp knitted and weft knitted fibres, knitted garments and hosiery. However in general the knitted and hosiery industry is fairly segmented and none of the other four leading firms (the Nottingham Manufacturing Company, Coats Patons, Carrington Viyella and Tootal) are represented in all these activities, but each has finishing capacity.

Although the degree of concentration in cotton and the hosiery and knitted sectors separately is not high compared with manufacturing industry as a whole, the concentration level in the combination of cotton, knitted and man-made fibres taken together is fairly high, as the leading firms are involved in more than one sector. Estimates made by the Textile Council for 1968 showed that the largest four firms in the cotton, knitted and man-made fibres sector (Courtaulds, Viyella, English Calico (now Tootal) and Carrington and Dewhurst) between them were responsible for 47 percent of spinning, 33 percent of weaving, 64 percent of warp knitting and 22 percent of weft knitting. In the finishing sector they accounted for 20 percent of bleaching (excluding surgical dressings) 62 percent of dyeing output and 61 percent of all printing. (4) A fifth firm, Whitecroft Industrial Holdings, controlled roughly half of U.K. bleaching capacity. Since 1968 the only major change among the main producers has been the merger of Viyella with Carrington Dewhurst in 1970. Table 1.4 gives estimates of the share of total output of cotton and allied textiles and knitted fabrics accounted for by the three main producers in 1972, and highlights their dominant position, which includes a 52 percent share of finishing activity.

TABLE 1.4 ESTIMATED SHARES OF TOTAL U.K. OUTPUT OF COTTON AND ALLIED TEXTILES AND KNITTED FABRICS, (1972)

(percentages)

	Courtaulds	Carrington Viyella	Tootal	Three firm total
Spinning	35	12	8	55
Weaving	25	15	6	46
Finishing	10	30	12	52
Warp Knitting	40	25	2	67
Weft Knitting	15-20	3	2	20-25

Sources: De Zoete and Bevan: The Major Textile Companies (1973) and industry sources

The woollen industry is less concentrated than these other sectors. The share of the largest firms showed little change in the 1963-68 period, in contrast to the cotton and knitted trades. Although concentration has grown more rapidly since then, reflecting the increasing shares of the leading firms (Coats Patons and Illingworth Morris), the structure is still relatively fragmented.

These concentration measures must be interpreted with care, as a high level of concentration does not necessarily indicate any lack of competition in the textile industry. This is because there is strong competition between the fibres, between alternative methods of fabric production and from imported yarns and fabrics, as well as increased concentration in the buying industries.

Number and size distribution

In 1973 there were 500 firms and 623 establishments with their major activity in textile finishing. There has been a substantial reduction in the number of firms, particularly between 1963 and 1968 when major structural changes were taking place throughout the textile industry. Between 1968 and 1972 the pace of decline slowed and the 1973 figures, the latest available, show a slight increase. The number of establishments has also declined in recent years, although changes in census definition may be partly responsible for this. Detailed information is also available about the number of works involved in the finishing of cotton and allied textiles from the Textile Statistics Bureau. At present returns are received from

TABLE 1.5 THE NUMBER OF FIRMS AND ESTABLISHMENTS IN TEXTILE FINISHING

Year	Enterprises	Establishments
1963	569	727
1968	468	622
1970	507	619
1971	476	615
1972	454	575
1973	500	623

Source: Census of Production

102 works engaged in bleaching, dyeing and finishing, 32 printers and 53 yarn processors, although the total number of works is less than these figures might suggest as some firms make returns for more than one activity. Table 1.6 shows the large number of finishing works which ceased production between 1963 and 1976 in cotton and man-made textiles (according to Textile Statistics Bureau returns) particularly before 1971, reflecting the decline and reorganisation of the cotton industry. During the period less than five finishing works were opened.

TABLE 1.6 FINISHING WORKS CEASING PRODUCTION IN COTTON AND MAN-MADE TEXTILES

1963-68	67
1969	11
1970	11
1971	19
1972	4
1973	4
1974	6
1975	7
1976	5
1977	4

Source: Textile Statistics Bureau

Size distribution of firms

In 1973 (the latest year for which the size distribution by establishment is available) the average establishment in textile finishing had 74 employees or a net output of £186,000. Average size has increased since 1968 if measured by net output but has fallen when measured by employment, as Table 1.7 shows.

TABLE 1.7 ANALYSIS OF ESTABLISHMENTS BY SIZE, 1973

All United Kingdom establishments classified to the textile finishing industry

Size Group	Establishments (Number)	Enterprises (Number)	Total Employment (Thousands)	Net Output Total (£ million)	Net Output Per Head (£)
1 - 10	205	201	1.1	32.5	2,572
11 - 99	278	258	5.6		
100 - 199	81	59	11.5	29.4	2,551
200 - 299	23	16	5.5	15.2	2,768
300 - 399	19	11	6.6	13.8	2,097
400 - 499	7	5	2.9	7.5	2,569
500 and over	10	9	6.9	17.5	2,528
TOTAL	623	500	46.1	116.0	2,515

Source: Census of Production

The most recent size distribution of enterprises is for 1972 (see Table 1.8). This shows that the average firm in textile finishing was fairly small with about 100 employees and a net output of £228,000. This can be compared with all textile activities where the average enterprise employed 154 people and produced £345,000 net output.

TABLE 1.8 ANALYSIS OF ENTERPRISES BY SIZE, 1972

All United Kingdom enterprises classified to the textile finishing industry

Size of Enterprise	Enterprises (Number)	Establishments (Number)	Total Employment (Thousands)	Net Output Total (£ million)	Net Output Per head (£)
1 - 99	366	395	9.9	22.2	2,242
100 - 199	51	58	7.1	15.3	2,168
200 - 499	22	31	6.1	16.3	2,647
500 - 999	9	30	5.8	12.6	2,165
1000 and over	6	61	16.1	37.1	2,304
TOTAL	454	595	45.1	103.6	2,298

Source: Census of Production

Economies of scale

The existence of economies of scale in the textile industry as a whole has been documented elsewhere (5), but information about cost conditions in textile finishing activities is rather fragmentary. The evidence which is available suggests that there are economies at the plant level in some areas of finishing, particularly bleaching and printing processes and the finishing of synthetic fibres. However these are generally not substantial and as there is also a need for variety in finishing processes, economies of scale cannot be considered a major deterrent to new firms wishing to enter the industry.

One measure of scale economies in textile finishing can be obtained by using the survivor technique. The principle of the survivor technique is to use the size distribution of plants (or firms) in the industry over time to separate the size groups which increased their share of activity from those showing increased shares. If a consistent pattern emerges, the size ranges with increasing shares can be identified as 'optimum' in terms of overall dynamic efficiency. Survivor estimates for textile finishing (and twenty nine other manufacturing industries in the U.K.) were obtained by Rees using 1958, 1963 and 1968 plant data. His analysis showed that the most efficient plant size was in the range between 100 and 499 employees and approximately one third of employment in textile finishing lay outside this optimum range. The number employed by a plant of efficient size was very small compared with total industry employment: 0.4 percent if one considers the average size within the optimum size class. These findings suggest that economies of scale are relatively unimportant for establishments primarily engaged in textile finishing and cannot be considered a source of entry barriers. (6)

The size distribution of firms and plants can also be used to see whether larger enterprises typically manage a larger number of plants than smaller enterprises and/or concentrate production in plants of larger size. In textile finishing both tendencies are apparent (see Table 1.8). The average size of establishments in terms of employment rises directly until a firm size of 500 employees is reached, suggesting plant scale economies exist, but as the size of enterprise rises throughout the distribution, so does the number of plants associated with them. This implies that there may be non-technical economies associated with firm size, such as preferential access to finance.

Among the main sources of scale economies at the plant level usually cited is the introduction of specialised machinery. This would tend to raise net output per employee until optimum scale was achieved (although there would be greater capital charges which are not reflected in net output). In 1973 the average net output per employee in textile finishing was £2,515. However productivity was highest for establishments with between 200 and 299 employees and showed no systematic tendency to increase with firm size, as Table 1.7 shows.

Another possible source of plant scale economies is the reduction in overhead costs through long runs. This is particularly important for commission finishers who do not own the materials they process, because costs such as wages and overheads, which are not directly dependent on output levels, form a large proportion of their total costs. The Census of Production shows that materials and fuel used accounted for only 38 percent of gross output in textile finishing, compared with 63 percent in the weaving of cotton, linen and man-made fibres, and 53 percent in hosiery and other knitted goods.

A further indication of economies of scale is the pricing policy adopted by sellers. Ewing (7) reports that the dyeing price list establishes a minimum of about 1800 yards for medium weight cloth above which no surcharges are made, but below which there are surcharges of up to 58 percent for quantities per shade of less than 250 yards. The printing list also offers steadily lower prices for long runs up to a maximum of 50,000 yards, depending on the fabric concerned and the type of order, the overall average reduction being from 40 to 45 percent, and in bleaching concessions for bulk orders are individually negotiated.

Specialisation and diversification

Many companies engaged in textile finishing concentrate almost entirely on the production and sale of textiles. The man-made fibre producers are the most diversified and ICI derives less than ten percent of turnover from textile interests. However the larger companies traditionally engaged in textile production have tended to diversify their activities within the textile industry rather than moving outside it. For example,companies in declining sectors such as woven goods have moved into relatively expanding areas like the production of knitted goods, with marginal adjustments to their finishing plant.

Smaller textile finishing firms tend to specialise in particular products and processes both because of the advantages of long runs and the need, in some cases, for specialised finishing machinery. The main area of specialisation in the knitwear and hosiery sector is fabric processing, and of the firms included in Table 9, eleven process knitted fabrics only. There are also specialised dyeing and finishing trades within the woollen industry of which mohair is perhaps the best example. The majority of the members of the British Textile Employers Association (which embraces the Lancashire sector) consider there to be only two or three competitors in their line of activity; the printers tend to specialise in three or four fabrics and dyers typically dye either synthetics or cotton and concentrate on a particular cloth such as velvet or corduroy. For example in 1976 there were only two indigo dyers in the U.K. equipped to dye the warps used for the denim trade.

Vertical integration*

Traditionally the textile industry had a predominantly horizontal structure, except in woollen spinning and weaving. Textile finishing operated as a distinct industry with specialist firms carrying out the different finishing processes, often on a commission basis. However, since 1960 the increasing importance of large vertically integrated groups has meant that many textile finishers have lost their separate identity and become part of a more closely linked textile industry.

---

* Vertical integration refers to the control under the same management of successive stages of production. Complete vertical integration in textile manufacturing would occur if a firm supplied exactly enough yarn for all its weaving or knitting requirements and had the dyeing and finishing equipment to process the fabric it produced.

The largest firms in the industry now all have vertically integrated structures, although they do not necessarily operate as integrated concerns. Courtaulds' share of activity diminishes at successive stages nearer the final consumer and output is least in finishing and making up. Work is sent to independent finishers on commission and some specialised finishing processes are carried out by Courtaulds for other firms. The structure of Tootal is the opposite of Courtaulds and as capacity in finishing is greater than in weaving and knitting, work is taken on commission to supplement that provided by the parent company. Carrington Viyella has a smaller variety of final products and most are processed within the group. The largest concern in the wool industry, Illingworth Morris,has interests in all aspects of wool manufacture but has only recently rationalised its structure to achieve the benefits of vertical integration.

The move towards vertical integration is well documented for the cotton and allied textile sector. Between 1961 and 1968 there was a very substantial shift from horizontal to vertical production. The greatest changes were in finishing activity where the capacity of specialist firms declined by 33 percent (8). Vertical integration did increase between 1972 and 1976 in all activities except yarn finishing but the changes were less rapid than in the previous period. By 1976, 232 firms (82 percent of the total number producing cotton and allied textiles) specialised in one activity. Of these 35 were specialist yarn finishers and 38 were cloth finishers employing 31 percent of those engaged in cloth finishing. 13 large integrated firms accounted for 77 percent of spinning employment, 52 percent in waste spinning, 44 percent of doubling, 56 percent of weaving, 61 percent of cloth finishing and they employed about 38 percent of those engaged in yarn finishing.

Vertical integration has also increased in the knitted and hosiery trades, although this trend has slowed recently. Table 1.9 shows the number of commission firms and integrated concerns in different size classes and throughput categories in 1976. It can be seen that in general commission finishers are still more important than integrated firms but their relative positions vary according to the product category. For example, in knitwear where stenters (dry setting finishing machines) are used, commission finishers with specialised equipment are particularly important.

In the worsted industry most organisations still confine their activity to one process but there has been a slow trend towards vertical integration here also. Vertically integrated concerns are much more common in the woollen system than in the worsted trade. This is because worsted cloth requires a number of separate finishing processes, the finishing plant is more complex and less adaptable than for woollen goods, and a worsted manufacturer is unlikely to keep all his finishing plant and labour utilised without causing delays. In 1968 it was estimated that the integrated firms handled about 70 percent of total woven cloth finishing but only 47.5 percent of worsted cloth (9). No detailed analysis has been made of the proportion of products dyed on commission, but estimates for 1968 show approximately equal numbers of commission dyers and spinning or weaving establishments with integrated dyeing facilities (10).

TABLE ANALYSIS OF FIRMS ENGAGED IN FINISHING HOSIERY AND KNITTED TEXTILES 1976 (a)

		Commission	Vertical
1	By number of employees		
	1 - 100	9	9
	101 - 250	11	4
	251 - 500	0	5
	501 - 750	2	0
	751 - 1000	0	0
	1000 +	0	1
	Total	22	19
2	By throughput category (b)		
	Ladies'hose and tights	4	7
	Half hose, three quarter hose and socks	7	8
	Fashioned garments and body pieces	5	8
	Circular knitted fabrics	15	11
	Yarn for hand and machine knitting	8	6
	Total	39	40

(a) The firms are members of the Federation plus two non-member verticals

(b) Most firms are engaged in more than one throughput category.

Source: The Knitted Textile Dyer's Federation

Mergers

Unfortunately separate data showing mergers in textile finishing are not available, but the experience of the whole industry acts as a useful guide. The initial shift towards a more vertically integrated structure in the textile industry was almost entirely due to large firms taking over some of the larger specialist companies as well as merging with other vertical groups. The take over of Bradford Dyers' Association by Viyella in 1964 marked the first time that a textile manufacturer became involved in dyeing and finishing on a very large scale. The man-made fibre producers played a particularly important role in the merger activity, partly as a way of securing their markets for man-made fibres. The Lancashire cotton industry, a major market, was especially threatened in the early 1960's by imports and competition in export markets as it was horizontally organised, very fragmented and lacked modern equipment.

Over the five years between 1963 and 1969, when the Monopolies Commission banned further mergers between the large firms, Courtaulds spent nearly £150 million on acquisitions in the cotton and knitted sector. ICI also invested substantial sums which were generally used to obtain minority interests in other companies, apart from the acquisition of a majority stake in Carrington Viyella in 1970.

Between 1958 and 1968 the textile industry was one of the most merger active manufacturing industries when mergers within the same industry are considered.* Although the Monopolies Commission prohibited further mergers between the five large firms in 1969, some mergers have taken place since, as Table 1.10 shows. However few major acquisitions have occurred and most transactions involve the building up of minority stakes in certain companies.

TABLE 1.10 EXPENDITURE ON ACQUISITIONS AND MERGERS OF INDEPENDENT COMPANIES IN THE U.K. TEXTILE INDUSTRY

	1970	1971	1972	1973	1974	1975	1976	1977 (first two quarters)
No. acquiring	22	22	43	42	18	6	11	11
No. acquired	30	37	52	50	21	7	15	12
£m (a)	21.2	17.7	30.5	31.6	11.3	2.4	5.1	10.2
£m as percent of total in manufacturing	3.6	5.9	2.5	8.4	6.8	1.9	2.9	5.3

(a) Valuation is by the market value of the smaller company's share of the newly formed company.

Source: Census of Production

Overseas trade

The market for textile products in Britain has only expanded slowly. Import competition, primarily from the Far East and Mediterranean countries, has affected a growing part of the relatively static home market and lower cost producers have competed in Britain's export markets, which are mainly found in Western Europe and in Northern America. Overseas trade figures are not available for textile finishing as this is a processing rather than a producing industry, but Table 1.11 gives a general view of the recent performance of three main categories of textile products. It can be seen that imports of textiles have accounted for an increasing proportion of U.K. consumption, particularly of man-made fibre and all fibres, so modifying the effects of rising concentration among domestic producers.

Britain, in common with many other advanced countries, has taken measures to control textile imports to give some protection to the domestic industry. However the 1973 Gatt Multifibre Agreement, which covers most textile products, provided for a progressive liberalisation of import controls and defined strict conditions under

---

* There were 66 mergers between textile firms compared with 368 'within industry' mergers in all manufacturing between 1958 and 1968 in the sample of all quoted companies with assets of £0.5 million in 1961 considered in Department of Trade and Industry <u>A Survey of Mergers</u> 1958-1968, H.M.S.O. London (11)

which quantitative restrictions can be imposed or maintained. The E.E.C. countries negotiate as a single unit under this agreement and the system for sharing the quotas of imports among member countries gave Britain a 23.5 percent share of total community imports and restricted the growth rate for many items to ½ percent annually. A new multifibre agreement is currently being negotiated but it is unlikely that import restrictions will provide significant protection in the long term.

TABLE 1.11 INTERNATIONAL TRADE IN TEXTILES

	Man-made staple fibre		All yarn (a)		All fabric (b)	
	Trade Balance '000 tonnes	Import	Trade Balance '000 tonnes	Import % (k)	Trade Balance '000 tonnes	Import % (k)
1966	+ 24.6	15.7	+29.9	4.8	-323.5	27.8
1968	+ 61.7	18.8	+31.1	7.2	-466.5	35.3
1970	+106.2	15.8	+34.9	10.1	-272.8	30.0
1972	+114.3	25.9	+36.1	15.5	-398.9	39.7
1974	+ 90.1	34.9	+13.3	22.9	-464.0	45.1
1975	+ 64.8	34.1	-12.6	22.3	-499.0	n.a.

(a) Woollen and worsted, continuous filament, spun man-made fibres and cotton yarns and threads
(b) Woollen and worsted woven, man-made fibre woven, cotton woven, and knitted fabrics
(c) Imports expressed as a percentage of home consumption (i.e. home sales plus imports minus exports).

Source: NEDO Textile Trends 1966-75

Although the textile industry has exerted strong pressure on the British government to increase protection, only minor concessions of this type have been given by increasing import controls in selected areas. Instead schemes offering financial aid have gradually been introduced, such as the assistance offered to the woollen and worsted producers in 1973 (under the 1972 Industry Act) to encourage modernisation and improve their future competitive position.

Financial data

Conditions in the British textile industry were relatively buoyant in 1973 but have been depressed since, due to the combined effect of the general recession and high levels of textile imports. These adverse conditions were reflected in short time working, redundancies, mill closures and reduced profits. Although turnover among listed textile companies increased by 16 percent between 1973 and 1975 (compared with the 43 percent increase in all manufacturing) net income fell by £167,026,000 during this period to less than half the 1973 level, as Table 1.12 shows. This can be compared with an 8 percent rise in the net income of all manufacturing companies. By 1975 the average rate of return on net assets earned by quoted textile companies was only 9.1 percent and the rate of return on turnover had fallen to 4.7 percent. (as compared with higher ratios for all manufacturing of 17.6 percent and 6.5 percent respectively).

Although manufacturing activity is showing signs of recovery from the recession, persistent destocking by distributors of textile products and high import levels have continued to depress the demand for textiles produced in Britain. Aggregate pre-tax profits reported by 206 companies engaged in the British textile, clothing and allied trades in 1976 fell by nearly £113 million compared with the results published by the same firms in 1975 and six of the largest groups together contributed a drop of £105 million towards the lower earnings. (12).

TABLE 1.12 PERFORMANCE OF LISTED TEXTILE FIRMS

	Turnover (£million)	Net Income (£million)	Net Income / Turnover (%)	Net Income / Net assets (%)
1973	2945	329	11.2	21.2
1974	3228	288	8.9	17.8
1975	3416	162	4.7	9.1

Source: Department of Industry

In 1976 there were 82 parent quoted companies with textile finishing activities (13). Table 1.13 presents profiles of the recent performance of six of the major companies, derived from company accounts. The performance of the Nottingham Manufacturing Company reflects the relative prosperity of the knitted goods sector, whereas the lower rates of return earned by Illingworth Morris mirror the difficult conditions experienced in the woollen trade. Courtaulds, Coats Paton, Tootal and Carrington Viyella all have interests in both cottons and man-made fibres but only in Coats Paton, which has diversified substantially in recent years, were returns above the average in manufacturing.

TABLE 1.13 PERFORMANCE OF MAJOR TEXTILE COMPANIES 1976-77

	Rank Among Industrials	Turnover (£ million)	Net Income (£ million)	Net Income / Turnover (%)	Net Income / Capital (%)
Courtaulds	14	1,510	117	7.7	12.8
Coats Paton	64	615	84	13.6	27.7
Tootal	110	325	24	7.5	16.6
Carrington Viyella	132	278	19	6.7	14.0
Nottingham Manufacturing	310	91	12	13.1	28.2
Illingworth Morris	311	90	4	4.0	7.3

Sources: Times 1000, 1976-77
Company accounts

THE UNITED STATES TEXTILE FINISHING INDUSTRIES

For purposes of comparison this section outlines some aspects of the structure and performance of the textile industry, particularly finishing activity, in the United States.

The traditional sectors of the textile industry have declined in importance in the United States in recent decades, as in Britain. The most prosperous and fastest growing section of the industry is the production of knitted goods where output increased by 47 percent between 1967 and 1972. The output of man-made fibres also rose but the existence of excess capacity has depressed prices and profitability. Fabric finishing output showed a decrease of 7 percent during this five year period.

Since the Second World War, major structural changes have taken place in the industry. The larger firms have tended to diversify their activities mainly into the faster growing sectors of the textile industry. They also invested substantial funds in modernising their equipment particularly in the 1960s and the industry as a whole is estimated to have reduced its technologically outmoded capacity from 29 percent in 1962 to 17 percent by the end of 1968. (14). Both forward and backward integration have taken place, but most of the large firms have not integrated forwards beyond the finishing stage. Until 1968 much of the structural change was achieved through merger but the Federal Trade Commission's 1968 guidelines, which restricted acquisitions in the textile industry, have limited the external growth of the largest firms.

Despite merger activities the industry is generally very fragmented and in 1972 there were 9539 establishments engaged in the production of textile mill products with a total sales value of $28 billion. Burlington Industries, the largest textile firm with $1.7 billion net sales in 1972, held only 8 percent of the domestic market in 1969 (15). Other large integrated textile mills include J P Stevens, Deering Milliken (a family owned firm) and United Merchants and Manufacturers. Apart from these four firms no other textile firm with dyeing and finishing facilities had as much as $500 million net sales (see Table 1.14).

TABLE 1.14 PERFORMANCE OF THE LARGEST TEXTILE FIRMS WITH FINISHING INTERESTS IN THE UNITED STATES 1971-72

Company	Net Sales ($ mill)	Net Profit ($ mill)	Return on equity (%)	Return on investment (%)
INTEGRATED TEXTILE MILLS				
Burlington Industries	1727	40.1	5	4
J P Stevens	861	-	-	-
Merchants	738	16.4	6	4
Lowenstein	444	8.9	6	4
INTEGRATED KNITTING MILLS				
Texfi Industries	164	11.3	25	16
Guildford Mills	53	2.7	37	25
Stevecoknit	33	2.4	27	21
Edmos Corporation	32	2.5	14	12
COMMISSION DYER				
United Piece Dyeworks	30	3.9	13.2	23
YARN SPINNING AND DYEING				
Dixie Yarns	99	2.3	7	7
National Spinning Co	76	1.9	8	8
Standard Coosa Thatcher	53	1.7	6	5
Burkyarns Inc	22	1.3	24	15

Source: US Environmental Protection Agency

This fragmented structure is reflected in the low concentration ratios, and makes the textile industry particularly vulnerable to import competition. The only sector in which the leading firm holds a substantial share of the domestic market is the production of man-made fibres where Du Pont has a dominant position, although its market power is also lessened by pressure from overseas producers.

The Census of Manufactures for 1972, the most recent available, provides detailed information about the structure of various sectors of the textile industry engaged in textile finishing. Table 15 shows the size distribution of establishments and indicators of performance, 1958-72, are given in Table 1.16. These tables show

TABLE 1.15 SIZE DISTRIBUTIONS OF ESTABLISHMENTS IN SELECTED SECTORS OF THE UNITED STATES TEXTILE INDUSTRY, 1972

	Wool Weaving and Finishing		Fabric Knitting and Finishing		Cotton Finishing		Man-Made Fibre Finishing		Yarn and Stock Finishing	
Employment	Plants	Employees	Plants	Employees	Plants	Employees	Plants	Employees	Plants	Employees
1 - 9	66	0.2	186	0.9	57	0.3	40	0.2	31	0.1
10 - 99	83	3.6	493	18.0	84	3.3	126	5.4	50	2.2
100 - 499	43	10.5	210	47.8	39	8.4	79	17.2	32	8.9
500 - 999	5	5.2	26	18.0	14	13.8	12	12.7	3	n.a.
1000+	1	n.a.	4	5.4	2	n.a.	2	n.a.	-	-
Total	198	19.4	919	90	196	25.7	259	35.5	116	11.2

n.a. not available for disclosure reasons.

Source: Census of Manufactures, 1972

TABLE 1.16 STRUCTURE AND PERFORMANCE OF SELECTED SECTORS OF THE TEXTILE INDUSTRY IN THE UNITED STATES, 1958 - 1972

	Number of establishments		Number of employees ('000s)		Value added ($m)		Value of shipments ($m)		Capital expenditure ($m)	
Sector	1958	1972	1958	1972	1958	1972	1958	1972	1958	1972
All textile mill products	7860	7539	918.6	983.8	4869.5	11715.8	12,415.8	28,063.9	218.5	1127.4
Wool weaving and finishing mills	469	198	56.0	19.4	336.6	239.4	929.0	450.1	9.7	11.6
Fabric knitting and finishing	na	919	18.4	90.0	147.4	1353.0	487.2	3796.5	7.1	318.2
Cotton finishing	446	196	49.2	25.7	289.1	322.1	686.2	623.3	15.7	31.5
Man-made fibres and silk finishing	197	259	16.2	35.5	114.8	526.5	210.6	1373.1	3.6	49.0
Finishing of yarn and stocks	155	201	7.8	18.5	52.0	270.3	158.0	637.3	2.1	35.2

na not available

Source: Census of Manufactures, 1972

that although textile employment and net output increased between 1958 and 1972, the number of establishments engaged in the production of textile mill products fell. In cotton finishing and wool weaving and finishing, employment and net output declined as well as the number of establishments, in contrast to growth experienced in other types of finishing activity. In general the industry has a very fragmented structure with small plants predominating. Plants with less than 100 employees accounted for 77 percent of the total number of plants in textile finishing in Britain and in the United States this size class accounted for 75 percent of all plants in wool weaving and finishing, 74 percent in fabric knitting and finishing, 72 percent in cotton finishing, and for 64 percent of the plants engaged in man-made fibres and silk finishing.

THE TEXTILE FINISHING INDUSTRY IN THE E.E.C.

Many features of the market structure and recent development of the textile industry in the E.E.C. are similar to those of Britain and the United States. Man-made fibres have grown in importance and in 1971 accounted for 54 percent of the total consumption of textiles in Western Europe, compared with 20 percent in 1950. In the same period production of man-made fibres in West European increased by 484 percent and excess capacity has reduced prices and the profitability of producers (see Table 1.17)(16). Output of knitted goods also rose, while the traditional wool and cotton sectors suffered from both the influx of cheap imported goods and the growing popularity of man-made fibres and knitted products. These trends were reflected in the changing composition of output of the textile finishing industry.

The most concentrated sector of the textile industry is the production of man-made fibres as in Britain and the United States. For example (17) before ICI's patent expired in 1967, several European producers began producing polyesters under licence, including Rhone Poulenc in France, Hoechst in West Germany and AKU/Glanzstoff (now AKZO) in West Germany and the Netherlands. Each national submarket in Western Europe is at present dominated by one or two firms with a few other small firms, usually the subsidaries of large international companies such as ICI or Du Pont, providing significant competition. In 1974 eight firms accounted for nearly 90 percent of the West European market for polyester . AKZO , (which owns Glanzstoff), British Enkalon and Hoechst dominate, each producing about 20 percent of the output. ICI, Rhone Poulenc and Du Pont have about 15, 12 and 8 percent shares respectively and Montedison, Snia and Bayer also feature among the major firms.

Substantial structural changes have taken place in other sectors of the textile industry, particularly cotton production, in response to the development of man-made fibres and increasing international competition. The large firms have become more vertically integrated and diversified since the early 1960s, often through merger activity rather than by internal growth. Levels of concentration have tended to rise, although the largest firms still hold fairly small shares of the individual national submarkets. Recent estimates provided in studies of the textile industry sponsored by the European Economic Commission (18) show that in 1972 the largest four domestic producers of cotton in France were responsible for 25 percent of the country's cotton production, compared with four firm sales concentration ratios of 21 percent in Italy and 20 percent in Belgium. The wool industry was slightly more concentrated especially in Belgium, where the four largest firms had an estimated 36 percent share of national output, compared with France's concentration ratio of 29 percent. In Belgium the production of hosiery and knitted goods was very competitive, according to concentration estimates, as the four largest companies accounted for only 15 percent of sales.

TABLE 1.17 PERFORMANCE OF THE LARGEST LISTED COMPANIES IN VARIOUS EUROPEAN COUNTRIES

	Rank among European Industrial Companies	Turnover ($ mill)	Net Income ($ mill)
W.Germany			
Hoechst Ag	10	7930	357.6
Enka Glanzstoff Ag	255	729	(34.8)
Gerrit Van Delden & Co	609	305	N/A
France			
Dollfus-Mieg and Cie	416	441	(1.9)
Lainiere de Roubaix	444	416	15.2
S.A. Texunion	528	351	2.5
Belgium			
P.R.B. SA	1225	140	3.1
UCO NV	1336	127	N/A
Italy			
Montedison Spa	23	5172	(238.5)
Snia Viscosa	191	990	(7.1)
Netherlands			
Koninklijke Textielfabricken	1143	152	(7.9)
Gamma Holding NV	1230	139	3.9

Source: Dun and Bradstreet Europe's Largest Companies 1977

Comparative data concerning the structure of the textile finishing industry in the E.E.C. is provided by an industrial enquiry carried out in six member countries, which was published in 1969 (19). This shows that 1921 firms were primarily engaged in textile finishing of which 802 were in Italy, 505 were in France, 321 were located in West Germany and the remainder were found in Belgium and the Netherlands. In general these companies were fairly small, only eighteen firms employed more than one thousand people and

85 percent had less than a hundred employees. The largest four firms accounted for a greater percentage of textile finishing employment in France (37 percent) than in Italy and West Germany, where the largest firms had 22 percent and 15 percent of employment respectively (20). These concentration figures can be compared with the 24 percent share of textile finishing employment held by the largest four firms in Britain at the time.

The size distribution of textile finishing establishments in these E.E.C. countries for 1963 also reveals a very fragmented structure, with a predominance of small plants. There were altogether 2065 establishments of which about 84 percent had less than one hundred employees. Average plant size was greater in West Germany (about 880 employees) and France (690 employees) than in Italy or Britain where the typical plant had a workforce of about 610 people (21). Unfortunately more recent data using a common industrial classification are not available for purposes of comparison, but recent structural changes in the textile industry, particularly the tendency towards vertical integration, have modified some features of textile finishing activity in the E.E.C. However, it still seems relatively fragmented and, for example, in 1972 the three largest firms in textile finishing in West Germany accounted for only 13 percent of employment and 15 percent of turnover (22).

REFERENCES

1. Department of Industry Report on the Census of Production 1973 HMSO, London 1977

2. Department of Employment Gazette

3. F.Fishwick and R.B.Cornu. A Study of the Evolution of Concentration in the United Kingdom Textile Industry. Commission of the European Communities, October 1975

4. Textile Council Cotton and Allied Trades - A Report on Present Prospects and Future Performance, Manchester 1969

5. C.F. Pratten. Economies of Scale in Manufacturing Industry Department of Applied Economics Occasional Paper No.28, Cambridge University Press, 1971

6. Rees: The Survivor Technique and Optimum Plant Size in the U.K., Economica, 1973

7. A.F. Ewing. Planning and Policies in the Textile Finishing Industry, Bradford University Press, 1972

8. Textile Statistics Bureau, 1976

9. NEDO The Strategic Future of the Wool Textile Industry, London 1969

10. Ibid

11. Department of Trade and Industry, A Survey of Mergers 1958-1968, HMSO, London

12. Tattersalls Trade Review, Manchester 1976

13. Extel Services

14. D.W. Huffmire, Strategies of the United States Textile Industry in the Post World War II Period. Journal of Business Policy, Vol.3 No.1 1973

15. Rush Loring, What the Textile Industry Really Needs, Fortune October 1970

16. G.L. Berrini, The Prospects for Man-Made Fibres in Western Europe, Congress of European Chemical Marketing Association, October 1973

17. This example is taken from R.W. Shaw and S.A. Shaw, Patent Expiry and Competition in Polyester Fibres, Scottish Journal of Political Economy, June 1977

18. European Economic Commission, Etude Sur L'Evolution de la Concentration Dans Quelques Sous-Secteurs de l'Industrie due Textile en Belgique (November 1975), Etude Sur L'Evolution de la Concentration dans L'Industrie du Textile en France (November 1975), Studio Sull'Evoluzione della Concentrazione Nell 'Industria Cotoniera Italiana (November 1975)

19. Office Statistique des Communautes Europeenes, Etudes et Enquêtes Statistiques, 2, 1969

20. K.D. George and T.S. Ward, The Structure of Industry in the E.E.C. - an International Comparison. Department of Applied Economics Occasional Paper No.43, Cambridge University Press, 1975

21. Ibid.

22. L. Phlips The Measurement of Industrial Concentration - A Reassessment based on European Data, Eurostat 3, 1975

Chapter II

Textile Finishing: An Overall View of Processes

The textile finishing industry uses large quantities of water and produces large volumes of aqueous effluent. The provision of water and disposal of effluents cannot be viewed as two separate problems as the viability of the industry in the future may well depend on the extent to which water can be reused. One important point is that each pollution problem is a separate entity and requires a custom-made solution. Any combination of unit operations (such as scouring, bleaching, dyeing and mercerizing) may be carried out at any particular mill and each unit operation generates a waste with its own particular characteristic. Also, each of the operations may be carried out batchwise or continuously, in kiers, jigs or other machines and to variable end points. Thus only generalizations can be made regarding the character of textile finishing wastes.

Dissolved and suspended matter in textile effluents arises from three sources; the textile material, the water supply and chemicals used in processing. Four waste producing variables are the amount of fabric passed through each process, water consumption, the amount of chemicals used, and the pollution contribution of impurities removed and chemicals used. In-plant measures to reduce the waste load should obviously be considered as an important part of any overall effluent treatment strategy. These can include the reduction in the volume ratio which will probably leave the waste load unchanged but increase its concentration. The reduction of waste water volume by counter-flow processing and re-use techniques (lowering water costs and increasing the concentration of effluents), the reduction of amounts of process chemicals used, the recovery of process chemicals (e.g. caustic soda in cotton processing, wool grease from wool scouring), process modifications (continuous operation is generally beneficial from a pollution point of view), substitution of process chemicals and "good housekeeping" (preventing spillages etc) may all contribute to a reduction in effluent disposal problems.

The determination of effluent characteristics can present serious problems since wastes vary in flow and strength to such a great extent. Flows are best monitored continuously by installation of a notch weir or similar device. There are various measures of the pollution potential or strength of a waste, and both a maximum and an average value should be known. The BOD test (usually taken over five days) measures the amount of organic matter that can be oxidized by micro-organisms in the presence of oxygen, which indicates either the load on a sewage disposal system or, in the absence of any treatment, the extent of de-oxygenation of a river or stream. Alternatively, the Chemical Oxygen Demand (COD) may be taken as a measure of the extent to which an effluent will deoxygenate a water course - the organic material is oxydized chemically rather than biologically. BOD and COD values are related but the ratio between BOD and COD will vary according to different effluent types. Immediate oxygen demand is measured by determining the Permanganate Value (PV); that is: the consumption of oxygen by acidified potassium permanganate. In the case of textile wastes, values of PV and BOD may not be related. Other measures of pollution load are suspended and dissolved solids, pH (measure of acidity or alkalinity) and the extent of toxic or otherwise damaging constituents such as sulphides, chromates, surface active agents, chlorine and hydrogen peroxide.

Treatment of textile wastes involves neutralization, together with removal of suspended solids and oxidation of organic substances. Discharge may be to a municipal sewer, with or without pretreatment depending on consent conditions, or sometimes to a river after full treatment at the factory to comply with stringent consent conditions. It is always necessary to have mixing and storage facilities for equalization of the wastes - this will prevent shock loads on treatment systems and permit initial design of a smaller treatment plant than would be needed to cater for maximum loads as they occur. One stage in preliminary treatment is the removal of solid matter such as fibres and sand particles, either by settling or more usually by screening. Neutralization may also be done at this stage, generally by addition of acids or alkali from a pH control electrode system (the overall waste characteristic is alkaline pH). An alternative to neutralizing the main flow is to segregate highly alkaline or acidic wastes and deal with them at source; one example is the segregation of very alkaline kier (scouring) liquors and either recovering caustic soda or neutralizing possibly with carbon dioxide from waste flue gases. Another example is the separation of wool grease from wool scouring liquors which entails addition of an acid to crack the soap stabilized grease emulsion. Pure grease may first be recovered by centrifuging.

Chemical coagulation or flocculation is sometimes used to remove some soluble organic matter and light suspended solids. Coagulants used are alum, copperas (ferrous sulphate), lime, ferric chloride, polyelectrolytes and a wide variety of other chemicals. Flocculation can be very useful as a pretreatment stage removing large concentrations of organic materials and possibly toxic substances such as chromium which might interfere with a subsequent biological process. One disadvantage is that the process is expensive, as large amounts of inorganic flocculants are needed. This also results in the formation of large amounts of sludge, but it is possible that a large amount of inorganic matter will make filtration more attractive than sedimentation and result in increased efficiency of sludge removal.

There is a tendency now to use chemicals which are readily biodegradable and to treat wastes by bio-oxidation processes, purification resulting from the use of organic contaminants by micro-organisms in their metabolic processes; the products are carbon dioxide,water, nitrates and cell materials. The normal BOD: nitrogen: phosphorus ratio for optimum efficiency of aerobic biological processes is 100:5:2. If the effluent is not being treated together with domestic sewage, nutrients may be added to try and achieve this ratio; ammonium phosphate is commonly used. Aerobic biological oxidation can be achieved by the use of trickling filters, where active slimes grow on a solid matrix of coke or granite and oxidation occurs as the effluent percolates over the matrix, or an activated sludge plant, where an aqueous phase containing colonies of bacteria built up on sludge particles closely mixes with the effluent. Both result in the production of additional solids (0.8 pounds for every pound of BOD) and BOD removal of up to 95 percent can be achieved.

Anaerobic digestion may be useful to preliminary treatment of sludges and strong wastes such a kier liquors to remove loads in anaerobic treatment plants. Carbon dioxide and methane are formed. Primary sedimentation may be carried out before anaerobic processing or after chemical treatment such as flocculation. It is not often necessary before aerobic treatment because the solids are more or less inert and sedimentation does not reduce the organic load significantly. Secondary sedimentation following biological treatment involves removal of biological solids and in the case of the activated sludge process rapid return of part of the sludge to the aeration system. An alternative system (which is not in general use) is to remove solids by flotation which is much quicker and allows return of solids to the activated sludge aeration system much more rapidly and in better condition. Flotation is achieved by attaching fine bubbles to particles using compressed air.

A good deal of residual BOD after biological treatment comes from fine suspended matter, and if this can be removed there is a substantial improvement in the effluent. If water is to be discharged to a water course or reused in processing some kind of tertiary treatment may be necessary. Polishing is achieved by use of grass plots, slow sand filters, microstrainers and lagooning. The removal of residual solid matter is more difficult but is only required for the removal of specific substances such as residual dyes or proofing agents. Small amounts of dyes are sometimes removed by activated carbon treatment.

The soluble organic impurities in the waste are converted in biological processes to solids, and finally rejected as a sludge containing 1-3 percent of insoluble matter. The bulk of water is generally removed by a combination of drainage and evaporation in drying beds, but this is a slow process. A thickening tank can first be used which speeds up subsequent drying. Dewatering may be carried out by vacuum or pressure filtration and in large works filter cakes can be brought down to inorganic constituents by incineration in multiple hearth or fluid-bed furnaces.

It is estimated that by the recycling of water and careful selection of rinse waters (e.g. rinse water from bleaching may be used for mixing after scouring) up to 40 percent in total water usage can be achieved. Biologically or chemically treated water may be suitable for reuse in processing or as cooling-water make-up. Some work has been done on the reuse of treated effluents for soil irrigation -

satisfactory results have been obtained by irrigation of grass plots using thickened sludge from aerated lagoons.

TABLE 2.1 EFFICIENCIES OF A SELECTION OF METHODS FOR THE TREATMENT OF TEXTILE FINISHING EFFLUENTS

Removal method	BOD %	Suspended Solids %	Total Dissolved Solids %
Screening	0- 5	5-20	0
Plain sedimentation	5-15	15-60	0
Chemical precipitatonn	25-60	30-90	0-50
Trickling filter	40-85	80-90	0-30
Activated sludge	70-95	85-95	0-40
Lagoon	30-80	30-80	0-40
Aerated lagoon	50-95	50-95	0-40

Source: FWPCA "The Cost of Clean Water" Vol.III Industrial Waste Profile No.4 Textile Mill Products September 1967

If substitutes for major process chemicals (e.g. starch, soap, acetic acid) causing high BOD loads are used, BOD reduction of up to 75 percent may be achieved. However, care must be taken since a substance with a low five-day BOD may have a high twenty-day BOD and thus cause deoxygenation some time after discharge. Also, the presence of non-biodegradable substances may be unacceptable, for example causing excessive foaming. This has been a problem in the increasing substitution of synthetic detergents for soaps - the latter are precipitated by calcium and magnesium ions in the waste water, but branched chain ('hard') detergents are resistent to degradation during sewage treatment, and can cause foaming in very low concentrations. The tendency now is to use straight-chain ('soft') anionic detergents and the concentration in rivers has fallen sharply, but the presence of small amounts of non-ionic detergents has a synergistic effect in stabilizing foam which would not persist in the presence of the anionic detergent only. Also, many auxiliary agents used in the textile industry have properties very similar to that of 'hard' non-ionic and anionic detergents, and alternative 'soft' agents should be carefully selected.

Although air pollution problems are secondary, some degree of control may be needed. One unit process giving rise to emissions is carbonizing, where vegetable matter or impurities in wool fabric are removed by sulphuric acid treatment, heating and mechanical dusting. Fine carbon particles are generated as smoke, as well as some fumes and odours, probably including sulphur dioxide as well as organic decomposition products. Fumes are usually exhausted to atmosphere, sometimes through a cyclone collector, but more stringent control may be required in the future. Possible solutions could be a high vent stack or wet scrubbing.

Odours of acetic acid, formaldehyde, acids and other organics may exist in higher than acceptable threshold concentrations and require control. Control methods include masking with another chemical, combustion, wet scrubbing and absorption on activated carbon. The use of an afterburner is probably the most efficient way of dealing with stenter fumes, since, although capital costs of such a plant are relatively high, the hot air can be recycled to the stenter with a resulting reduction in energy costs. Alternatively, control may be achieved through change of process chemical or design modifications (enclosures and hoods). High vent stacks may be utilized to reduce ground concentrations to below the acceptable threshold.

Chapter III

Wool Dyeing and Finishing

INTRODUCTION

This industry is mostly situated in Yorkshire, and has potentially the severest pollution problem of the three main sectors of the textile industry. Our five cases give a very good coverage of the whole industry.

(i) PRODUCTION PROCESSES AND CONTROLS

The conversion of raw wool into useful material involves several stages of production, some or many of which will be carried out in separate firms so, even though we have identified here firms engaged in the dyeing and finishing of wool, it is inevitable that there is still a wide range of processes used by these firms. Raw wool is first scoured and may then be converted into yarn; after it has been converted into yarn, it may be dyed as yarn or the undyed yarn may be converted into cloth which is then treated in various ways which includes dyeing. Some firms may treat the raw wool and convert it into yarn; whilst others may dye the yarn as well. Other firms may deal with only the dyeing and finishing of cloth. Thus, of the firms in our sample, one is involved in the scouring of raw wool and its spinning and dyeing; another is engaged solely in dyeing of wool yarns; and the other three are engaged in dyeing and finishing of fabric.

The diagram shown in Figure 3.1 illustrates the processes through which the fabric passes after the raw wool has been scoured, spun and woven into fabric. Some of these processes are also used in the dyeing of wool yarn after the scouring and spinning stages. Wool contains natural impurities such as grease (insoluble in water), suint (water-soluble); and added impurities such as dirt and paint. These impurities are largely removed during the scouring of the raw wool and these scouring liquors contain a high proportion of emulsified wool grease and other substances and they have very high BOD loadings. However, the scouring liquors from wool *yarn* and *fabric* contain much less grease but also contain spinning lubricants applied during processing, together with soap or detergent.

FIGURE 3.1

WOOL FINISHING WASTES

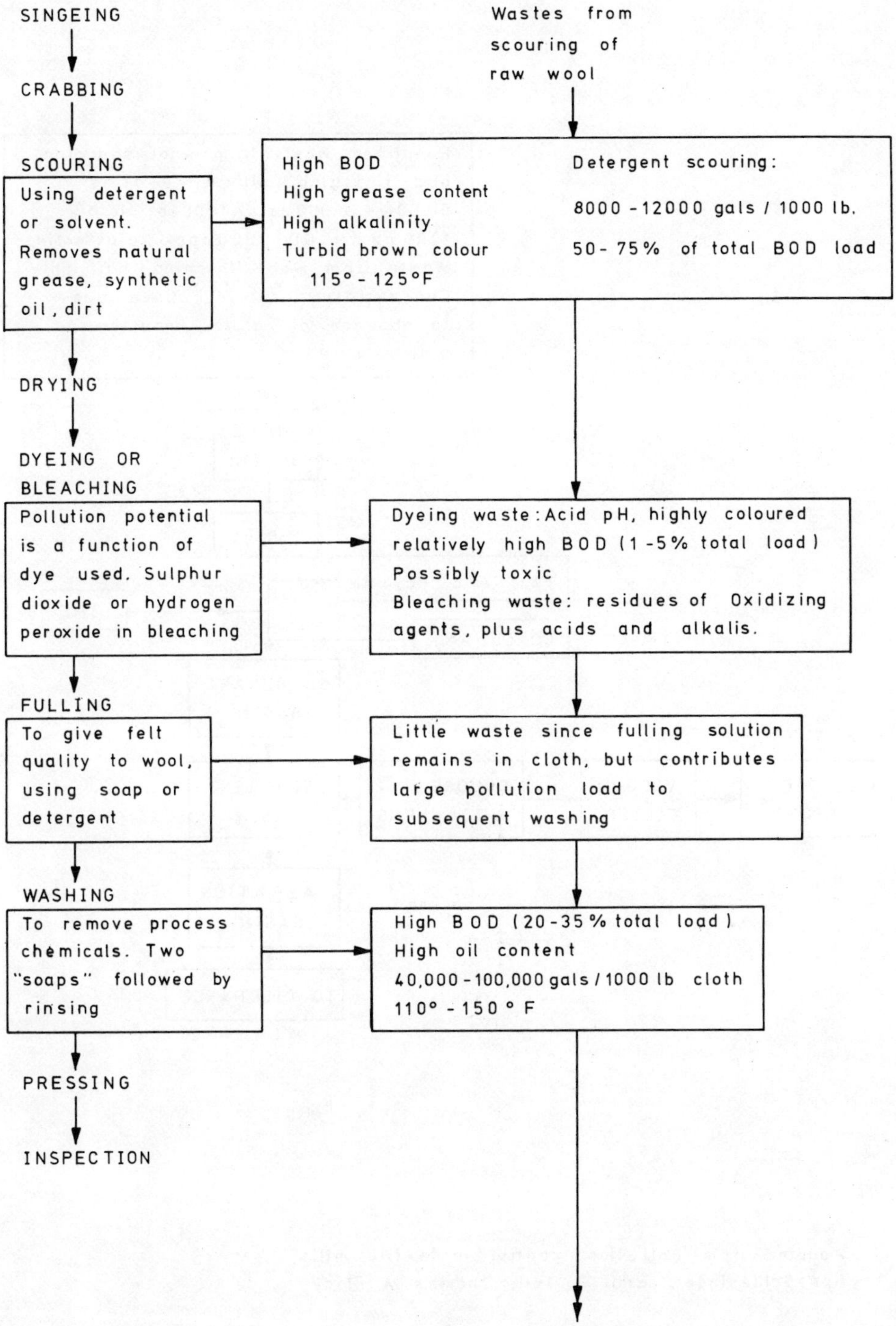

CONTINUED

FIGURE 3.1 contd.

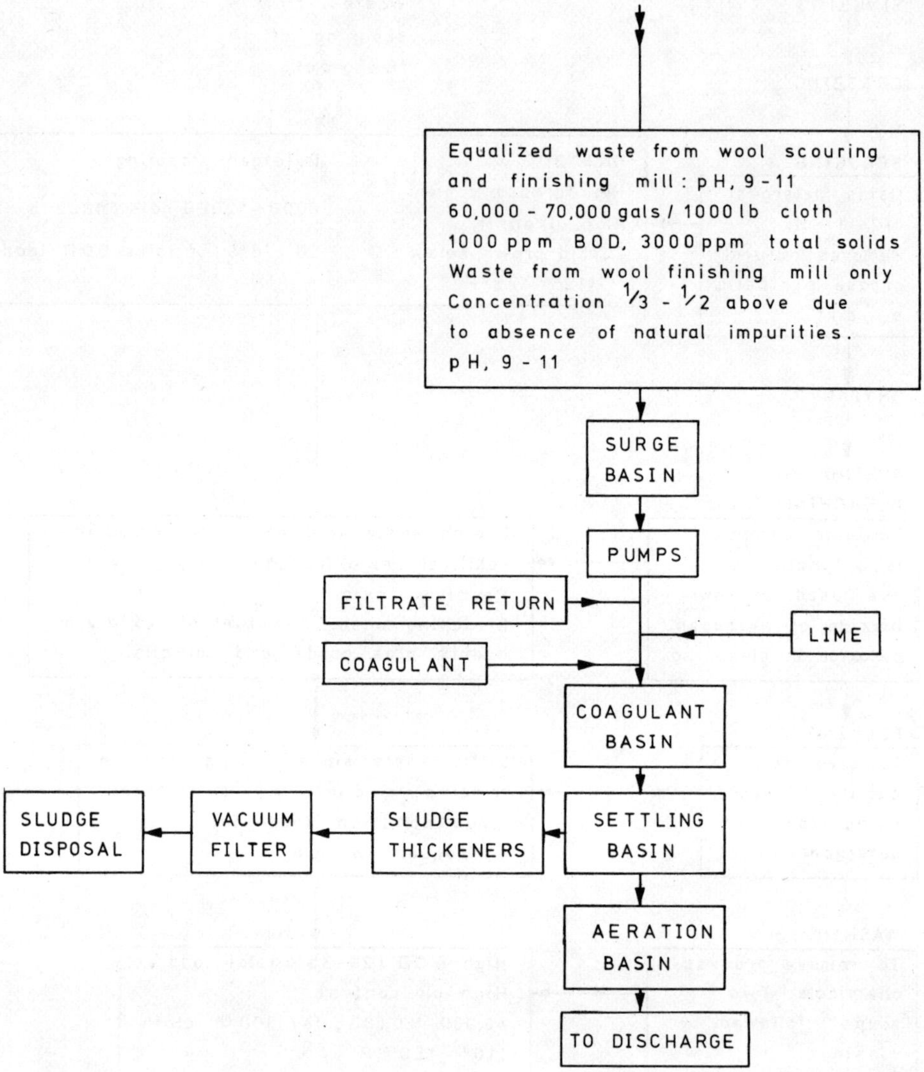

Adapted from "Pollution control in textile mills"
H. A. Schlesinger, Emil F Dull, Thomas A Fridy

After possible singeing and crabbing, the fabric is scoured. This process involves first steeping, and then scouring in soap and alkali (usually sodium carbonate). The alkali saponifies fatty acids in the wool grease to form natural soaps and the soap emulsifies wool fat, forming suspensions of small dirt particles. More recently synthetic detergents have been used instead of soap; this results in less damage to the wool but little overall change in the pollution potential. Alternatively, a solvent scouring technique may be used; this produces a lower pollution load but must be followed by a detergent wash to remove dirt. It can be performed batchwise or continuously in a kier and the grease-laden spent solvent is recovered, redistilled and reused. Suint and traces of solvent have to be removed by washing. Further, these solvent scour methods find increasing difficulty in meeting tightening standards for *factory* atmospheres.

Wool grease may be recovered from waste water by two methods: centrifuging which recovers the grease in a relatively pure form or acid-cracking which is used for the recovery of fairly impure grease. The wool grease is a source of lanolin; the potassium salts could be recovered from suint but the main drawback to this is the lack of a favourable market. If both lanolin and suint are recovered, it is estimated that a 50 percent reduction in the BOD loading of a scouring and finishing plant could be achieved.

After rinsing and possible drying, the fabric may be dyed or bleached. The BOD loadings in dye waste-waters can vary a great deal depending on the types of dye used. For example, acetic acid dyeing produces a waste with around 2,400 ppm BOD but ammonium sulphate dyeing produces a waste with around 600 ppm BOD. The waste waters are highly coloured and may be toxic; for example, sulphur dyes are toxic in themselves and high concentrations of dichromate ions can be damaging to biological treatment. In the latter case, the dye waste may be segregated and the dichromate reduced and precipitated before it reaches the waste treatment plant. The colour is a problem from the aesthetic point of view and the colour of many newer dyes can only be reduced by 50 to 85 percent by activated sludge treatment and it may be necessary to adopt other methods such as polymer coagulation, carbon absorption, or ion exchange. Generally the practical solution is to dose with hypochlorite or ozone in lagoons before further treatment. In intractable cases an alternative dye selection may be necessary.

If white or light shades of coloured cloth are required, the wool is first bleached. Hydrogen peroxide and sulphur dioxide may be used and their contribution to the total BOD load is slight, about 0.5 percent of the total; optical brightners may also be used and these may contribute 1 percent of total BOD load.

After dyeing or bleaching the fabric may be fulled to give it a felt-like quality; this process uses soap or detergent. The fabric is then washed and this washing process is the second largest source of BOD load. The BOD may range between 4,000 and 11,500 ppm and total solids range between 4,800 and 19,300 ppm. The wool fabric is then carbonized by impregnating it with 4-6 percent sulphuric acid and drying it in an oven at a temperature between 212° and 220° F. Water is evaporated off and the remaining organic matter is charred; the charred matter is then crushed and knocked out in a mechanical dusting machine. The fabric is then rinsed in water and passed through a bath of 2½ percent sodium carbonate for neutralisation of residual acid. The waste water from this stage has a pH range between 1.9 and 9.0, total solids in the range 1,200 to 4,800 ppm and may contribute about 1 percent to the total BOD load of the final effluent. The cloth is then washed again and dried.

The fumes from the carbonizing process are collected by means of stainless steel or plastic coated hooding and exhausted to the atmosphere, sometimes passing through a dry cyclone collector. It is possible that more efficient removal of carbon particles and a reduction of sulphur oxide emissions may be required which could be achieved by venting the exhaust to the main boiler stack or by installing wet-scrubbing or dry-filtration facilities.

The "typical" characteristics of the liquid effluent from a wool scouring and finishing mill would be as follows: pH 9 - 11, 60 - 70,000 gallons/1,000 lbs of cloth treated, 1,000 ppm BOD and 3,000 ppm total solids. The waste from a "typical" wool finishing mill only would have the same pH value but concentrations about one third or one half of these figures due to the absence of natural impurities. In the past the most common method for treating these wastes was chemical coagulation, with or without pretreatment for grease removal; these methods achieved control efficiencies between 20 and 85 percent. The trend now is towards biological treatment where necessary, modified activated sludge treatment with influent pH adjustment can achieve control efficiencies of the order of 90 percent. When detergent scouring is practiced, biological treatment is preceded by a chemical coagulation stage to reduce BOD loadings but the need for this step is removed by the use of solvent scouring which results in little or no grease passing into the effluent.

(ii) CASE STUDIES

Case 1

This large firm employs over 1,500 people and processes a significant part of all the raw wool imported into this country from Australia. It is involved in the scouring of raw wool and spinning and dyeing of wool and synthetics. Throughput is roughly 100 tonnes of wool and a similar amount of synthetic material per 120 hour week.

The plant was originally located on its present site because of the availability of a suitable water supply from a borehole, but this now has to be supplemented with town water; approximately 2000m³ of water are used every 24 hours and this is accounted for as follows:-

dyeing and yarn scouring	1600m³
raw wool scouring	180m³
lost as steam	220m³

Roughly one half of all water comes from the firm's borehole. All water is heated before it can be used in processing. After treatment, the effluent is discharged to a municipal sewer and the consent conditions attached to this discharge since 1st April 1976 are:

grease	800 ppm
suspended solids	1000 ppm
maximum discharge rate	113.5m³/hr
maximum discharge volume	2000 m³/day

These conditions have been relaxed slightly to bring the effluent discharged to well within the limits. In the past the firm has paid a flat rate of £30,000 per annum as effluent charge but this was expected to increase, but the firm was unable to give any indication of the magnitude of any likely change.

The highest pollution load originates with the scouring of raw wool and, although it is technically not part of the finishing process, we include it since this is an integrated factory conducting both scouring and finishing operations. Raw wool is scoured at a temperature of 45°C with soap and alkali in order to remove the grease and other impurities; the untreated effluent contains 36,000 ppm of grease, together with 50,000 ppm of other suspended solids. This, together with its alkalinity (it has a pH of 9) makes it highly polluting and direct discharge would not be permitted. Although it accounts for roughly 90 percent of the process water, the dyehouse effluent is less potentially polluting; it contains 1000 ppm of grease and 5000 ppm of suspended solids, is highly acidic (pH 4) and its temperature is high, often quite close to 100°C. The company said that none of the chemicals used in the dyehouse is toxic. Yarn washing produces a fairly clean effluent which contains large amounts of fibres which are difficult to remove since screening is not possible as any fine mesh becomes blocked very quickly. Substantial control measures are needed to bring the combined effluent up to a standard to meet the consent conditions.

Some measures have been taken in recent years to lower the pollution load of effluents at source,however, the firm was unable to supply any firm quantitative information and many potential savings have not yet materialised. Soap and alkali for scouring is now metered automatically and this could theoretically reduce chemical usage by 30 percent through factors such as prevention of spillages but savings of this magnitude have not yet been apparent. Reduction in the amounts of chemicals used would have the effect of lowering the pollution load of the effluent. Further, £1 million was spent on the installation of a new scouring system utilizing one bowl instead of three which was potentially capable of reducing water consumption (and hence the effluent volume) by 30 percent. A heat exchanger installed in 1969 at a capital cost of £30,000 reduces the temperature of the dyehouse effluent to a level within the consent conditions and also heats the incoming water. Since there are strict limits on the temperature of final effluents, this must certainly be viewed as pollution control equipment even though there are substantial energy savings arising from it; 4,500m³ of water are heated by 20°C every week and the plant has quickly paid for itself.

The method of removing grease from the effluent from wool scouring may be viewed as "in plant" pollution control since the grease is recovered; first nearly one third of the grease is recovered by centrifuging and this yields pure lanolin. Since this can be a highly profitable operation in its own right - the lanolin is sold for up to £500 per ton - the firm would almost certainly carry out this operation even if there were no pollution controls. The price of lanolin can fluctuate, of course, but even at a price of £300 per ton it would be profitable to recover the lanolin by centrifuge. Ninety percent of the remaining grease is recovered by acid-cracking and this must definitely be considered as pollution control since the recovered grease is impure and fetches a lower price which is probably not sufficient to cover operating costs. The cracking plant is old and not much expenditure has been devoted to it in recent years but its replacement value is estimated at a quarter of a million pounds; labour costs for this plant are estimated at £20,000 p.a. and maintenance costs are also high.

After acid-cracking the scouring effluent has an acidic pH of 2 and it must therefore be neutralised. Lime and ferrous sulphate are added to bring the pH back to 9 and to act as flocculants for chemical precipitation. The dyehouse effluent, at pH 4, is slightly coloured and is mixed with the scouring effluent to give a neutral resultant liquor at pH 7. Precipitation brings the pollution load of this effluent down to 200 ppm of grease and 200 ppm of suspended solids at a temperature less than 32°C and this is then discharged to a sewer. Formerly the sludge from precipitation was disposed of at a weekly cost of £1,000 but now it is returned to the acid-cracking plant and the cost of disposal has been reduced to £150 per week although the cost of running the cracking plant has been increased (by how much, we were not able to ascertain).

It is very difficult to estimate the true costs of pollution control of this firm; for one thing, much of the plant is now old and fully depreciated and the control equipment, as with other plant, has been installed in a piecemeal fashion rather than as a "one-off" item of capital expenditure. Apart from this, there are many "grey" areas where it is difficult to be categorical about whether particular parts of the plant are for pollution control or not. Some such items have already been mentioned; another example is the dust extraction system recently installed in the carding room at a cost of £250,000. Part of this should be allocated to pollution control since the new scouring process has the effect of leaving the fibres in a dustier condition. Other hidden costs of pollution control had also been incurred; over one hundred thousand pounds had been spent on research into a new solvent scouring process to remove the effluent problem but none had been installed; further research had also been undertaken into the techniques of ultrafiltration and reverse osmosis. These research costs should be added to the actual expenditures incurred by the firm on pollution control even though the techniques investigated are currently uneconomic.

In addition to the disposal of liquid effluents, the firm has an odour problem which occurs on a few days every year. The mill is situated close to residential housing and odours from the acid-cracking plant sometimes give rise to complaints during the summer. This could conceivably become a serious problem in the future although the firm is considering the installation of fans to drive the fumes up the chimney although this would generate a problem of noise.

We were surprised to find that a firm as large as this and with not inconsiderable pollution problems was unable to give us any detailed information relating to the costs of control - we gained the impression furthermore, that this was not due to any reluctance on its part. Nevertheless, we believe the pollution control costs (ignoring the true replacement costs of machinery) to be small; the gross pollution control cost is of the order of 3p per kilo of processed wool and the net cost may well be negative if the revenue from the sale of lanolin is included. The gross pollution cost for the firm is of the order of £150,000 per annum when effluent charges, sludge disposal charges, labour, chemicals and depreciation are included.

Value added is composed approximately of 16 percent in combing and scouring raw wool, 44 percent in spinning and the rest in dyeing. Pollution control accounts for about 4 percent of value added in dyeing. However, whilst it is recognised that these costs are low, in comparison with profitability levels, they may be very high and it is probable that the pollution control cost is close to what would be considered a good profit rate on wool combing.

<u>This large dyeing complex gains from economies of scale in that it can treat its own effluent and recover a useful cash product. To do this, however, requires considerable capital expenditure as is indicated. The relaxation of the consent conditions does not imply that either effluent quality from the firm or final treated effluent quality from sewage treatment works will be reduced.</u>

<u>The description extends the total concept of pollution to include other textile operations. These have to some extent escaped attention in current discussions.</u>

<u>Case 2</u>

This firm, part of a large group, employs about 110 people in a medium-sized dye-house carrying out commission dyeing of wool yarn; in addition, a very small amount of synthetics is also dyed. Most of the wool yarns require scouring before dyeing and it is this process which produces the strongest effluent.

Water usage is about 4,500m^3 per week and effluent is discharged at a constant rate of about 45m^3 per hour. The water is obtained from a borehole and softened by ion-exchange; at present, therefore, the supply of water costs nothing apart from the minimal licence fee but it is possible that an abstraction charge will be made in the future. The cost of treating the water is small and it is made up of costs of electricity, depreciation, maintenance and salt for the regeneration of the ion-exchange column. About 500m^3 of water are lost every week as a result of leakage from the mill dam.

Scouring is carried out continuously and is arranged so that the flow of scouring liquor runs counter to the flow of materials so that the dirtiest liquor comes into contact with the dirtiest material and as the material becomes cleaner so it is transferred to the cleaner scouring liquor. The effluent from scouring (which is about one-fifth of the total volume of effluent) has an alkaline pH, an average suspended solids content of 300 ppm and a very high average COD loading of 33,000 ppm. The dyeing liquor is much weaker although the total volume is high; the average COD loading is 850 ppm, suspended solids 25 ppm and the average pH is 4.. Before discharge, the effluent first passes through a heat-exchanger used to heat the incoming water. This was installed in 1967 at a cost of £10,000 and amortised itself in a period of 18 months; it is estimated that at present fuel prices the fuel saving is between 15 and 20 thousand pounds per annum. The firm is considering the installation of another heat exchanger to make the process even more thermally efficient; some heat loss is regarded by the firm as inevitable, however, and there is a temperature loss of about 14°C as the effluent passes through open culverts prior to mixing in the central storage tank and passing through the heat exchanger. All effluent is concentrated in one mixing tank and since this was installed, the *overall* COD load has been reduced. By mixing the two effluent streams the overall COD loading has been reduced to between 3,200 and 7,000 ppm although the firm still finds pH control difficult to achieve.

Until shortly before our visit the firm paid no effluent charges at all. However, the water authority has now begun to levy charges and the first two annual charges amounted to £10,000 each, reduced by 50 percent under transitional arrangements. We had earlier expected that charges would have been £50,000 p.a. but by concentrating the effluent in the one mixing tank the firm has - as already mentioned - substantially reduced overall COD loads and this has made a marked impact on the effluent charge. The reduced rate transitional arrangements are not expected to last much longer but the firm is hoping that a concession will be made enabling oxygen demand to be calculated as BOD rather than COD and it is felt that this could substantially reduce effluent charges further (the BOD load is reckoned to be about 80 percent of the COD load).

Given that the turnover of the firm is one and a quarter million pounds per annum, an effluent treatment charge of £10,000 p.a. would represent 1 percent of total costs including raw materials; this works out at about 0.25p per kilo of yarn treated.

The firm does not envisage that it would ever be justified in installing its own treatment plant although it is possible that plant for full treatment and recycling of water might become attractive in the future. This has been discussed with the water authority and the firm decided not to install its own facilities.

In addition to the charges themselves and the capital cost of the mixing tank for effluent, there will be other costs to be met; for example, three flow meters have just been installed at a cost of over £300 on the recommendations of WIRA. Further, the acidic effluent will have to be neutralized before discharge and there will be an ongoing cost for chemicals as well as a certain capital expenditure.

The firm is looking at alternative dyeing methods - at present liquor ratios are about 20:1 but new equipment can achieve liquor ratios of 2:1 or less. It is quite likely that there will be a gradual replacement of dyeing machines, resulting in a reduction of effluent volumes; at the same time there would be an increase in the strength of the effluent so it is not easy to predict the impact of effluent charges.

This dyehouse is a good example of what can and needs to be done when it is necessary. Eventually, in terms of current costs, an additional figure of not less than £25,000 will have to be carried.

The measures taken and the longer view - including the examination of low liquor-ratio dyeing plant - are right.

Case 3

This is an integrated firm with its own finishing department employing about 30 people; at the time of our visit it carried out piece scouring, wool top dyeing, its own piece dyeing and some commission dyeing, but subsequently it has ceased the wool top dyeing. Although the firm has some stenters, there are no fume problems since no phenolic carriers are used. The finishing department operates on a day shift only.

Water usage is a maximum of 3,600m^3 per week; until shortly before our visit this was extracted from a borehole and pretreated but because of a deterioration in the quality of this water, the firm has switched to using town water. This is more expensive than water from the borehole and there is not the guarantee of supply so the firm was still hoping to find some way of using the borehole water again. Motorway works close to the plant are believed to have caused the change in raw borehole water constitution, but this cannot be proved. The average quality of liquid effluent to the main sewer were as follows:- settlable solids 220 ppm, COD 550 ppm, pH 5 to 10 and chrome content less than 5 ppm. No pretreatment of effluent was carried out apart from settling, but the effluent easily came within the limits set. The main problem was that of volume and most efforts had been, and were likely to be, devoted to looking at ways of reducing the volume of effluent. Total water charges were £15,000 per annum in 1975 (including water and effluent disposal); about 40 percent of this was the effluent charge (3.9 p/m^3). However, with local government reorganisation the new water authority was expected to raise its charges fourfold over the next four years. The firm felt that this could make in-plant treatment and total recycling of water economical. Up to our visit, only one measure had been taken to reduce water usage, the least contaminated portion of the scouring rinse water being recycled back through a storage tank. The capital cost of this plant, together with associated pumps and other equipment was small but the total water saving was only of the order of 2.5 percent. The firm found that its usage of water in scouring was higher than the theoretical level but a solvent scouring process would not have been a suitable alternative since the soap alkali process also gives the desired quality to the worsted cloth. Dyeing also uses a large volume of water; old winch machines are used that could be replaced by new machines with lower liquor ratios but the capital cost would be very high and such an investment was not envisaged.

This case illustrates the difficulties that a small dyehouse may have to meet. A changing quality of supply and the prospect once again of increased charges. It is interesting to note the dichotomy of thought on capital expenditure.

A fourfold increase from 10p./m^3 certainly is (currently) at the point where recycling looks attractive: but it is very expensive of capital, no less so than new winches.

Case 4

This firm is probably one of the smaller branches of an integrated group. It acts entirely on a commission basis and employs about 95 people, handling approximately 500 pieces of cloth per week. Other dyeing firms within the group are more vertically integrated and tend to handle easy materials; short runs of more difficult materials are farmed out to commission works such as this. Each piece of cloth weighs about 60 lbs and 90 percent of the throughput is wool worsted suitings, 5 percent is ladies' dress goods and the remainder is miscellaneous materials. The firm deals with some "fancies" that do not go through the dyehouse but pass through other finishing operations. The respondent said that a long-term factor leading to a reduction in water consumption was the continuing tendency to light-weight suitings (as a result of central heating) which require less water than the heavy-weight suitings common some years ago.

The firm uses about 3,000m³ of water every week and this is approximately 4.5m³ per piece of cloth. 1,100m³ are softened by a base exchange process and heated to a temperature of 120°C for scouring, 350m³ are heated to boiling point for dyeing purposes and 450m³ are used for steam raising, the remainder is used as wash water and is not heated. Heating costs approximately 44p/m³; this includes the capital cost of the boiler-house (which is now virtually fully depreciated) as well as fuel, labour and maintenance. Effluent is discharged direct to the sewer; the consent conditions attaching to this are:

maximum flow	135m³/hour
maximum temperature	40°C
pH limit	6-10

The firm has no difficulty in meeting these limits. At the time of our visit the effluent charge was 2.6p/m³ but this had risen from a level of 0.4p only shortly before and the latest predictions indicate further substantial rises in the near future. The firm is disturbed by these increases since before the inception of the new water authority, the local authority had been considering a scheme that would have meant that the firm would have paid 11p/m³ for its effluent, but this was shelved. Given such a charge, the firm would have been under considerable pressure to reduce its throughput of effluent and, indeed, the firm believes that had it been aware of the rapid price rises that actually have taken place since the formation of the water authorities, its whole capital investment plans would have been different.

Three principal operations are carried out; scouring and washing, dyeing and final finishing. These processes account for weekly water flows of 1800, 550 and 140 m³ respectively. The characteristics of the effluent are pH 9.7, temperature 80°F, BOD 500mg/litre, COD 550mg/litre and settlable solids 300mg/litre. The firm has been involved in considerable sampling of its effluent characteristics and is concerned with the variability of the results obtained. Even the same laboratory using two samples divided from one sample can give BOD ratings that differ considerably and the firm is naturally concerned that its charges may in future be dependent on unreliable analytical techniques.

The firm has not found it possible to make in-plant changes to reduce pollution load by reduction in chemicals or by change of chemicals. The amount of heat recoverable from the effluent is considered very low and there would seem little benefit in this. The firm has, however, made a fuel saving of about 10 percent by switching work patterns in the wet business, now 4 shifts of 10 hours are worked instead of 5 shifts of 8 hours. Even higher savings could be achieved by continuous working. The firm has recently managed to save 220m^3 of water per week by re-using final scouring wash-water in the first scouring. This has cost £1,500 for pipework valves and catchment and storage tanks but it is anticipated that it will amortise itself quickly (the weekly water saving is £30). The firm considers that equipment manufacturers are making some moves towards this themselves by incorporating holding tanks into the equipment to enable clean water to be reused.

The firm carries out some pretreatment by screening its effluent; there are rough screens at the ends of the dyehouse and scouring drains and a fine screen at the catch-pit where these two sources combine. This results in a reduction of the BOD load of 50 or 60 percent due to the removal of flock. The screens cost very little indeed but the firm has also investigated other methods of control. One would have been a balancing tank and aeration plant; this would have cost £15,000 for a 75 percent reduction in BOD load.

A full biological treatment plant could have given a control level of 87 percent at a capital cost of £45,000. Both the latter figures are expressed in terms of 1974 prices.

The firm tips waste fibres although some is sold to hop farmers for a small sum. The firm has no air pollution problem; the oil-fired boilers are exhausted through a stack 35m high and there are no odours arising from the steam-heated tentering equipment.

This small woollen mill has had to absorb quite enormous effluent charges over a short period. The current figure of 11p/m^3 is already high and it would seem that some hard capital expenditure decisions will be needed once the charges are BOD/COD related.

It is interesting to speculate on the way in which capital investment plans would have been different if the way in which effluent charges were to rise had been known.

The value of counter-current use of water is again demonstrated.

Case 5

This is a small independent firm which shortly before our visit had to contract to one third of its original size in order to raise sufficient capital to continue operating. It is mainly involved in spinning and weaving but it has a small scouring department employing 2 people on average and also carries out some dry finishing operations such as raising.

Average water consumption is about 250m^3 per week; this is obtained from the river and does not require any treatment before use. There is also a metered supply of town water to the works for use in the canteen etc., but use of this for scouring is avoided as far as possible because of the cost involved. At the time of our visit, the cost of effluent treatment was 2.6p/m³ (roughly £300 per annum) but by 1977 it had risen to 3.4p (£380 per annum). It was expected that when a new water authority charging system was implemented, the cost would rise fourfold and so the firm had become concerned with effluent volume and quality.

The scouring process consists essentially of three stages. Initially large amounts of detergent and soda ash are added to remove most impurities and the resulting dirty water is run-off. More water and detergent is then added for a period of 20 minutes and finally there is a long wash-off period giving rise to a large volume of much cleaner effluent. The firm had thought of the possibility of re-using this wash-off water which would, it had been estimated, reduce water consumption by thirty percent. The reduced effluent volume would mean that total charges would be reduced but the precise extent of the reduction would depend upon the strength of the resulting effluent. The project had been postponed since the capital cost of storage tanks, pumps, piping and other necessary work would have been quite high and both capital and space imposed severe constraints upon the firm. In fact, even in late 1977 the firm told us that the cost of water saving projects still could not be justified on the basis of return on capital, even allowing for increased consumption of water following an upturn in trade.

The firm once considered the possibility of closing its own scouring department and sending the work out to other processors, but this would have proved uneconomic. Any radical change in scouring process - such as a change to solvent scouring - was also out of the question because of capital constraints and because it could not give the required quality of finish. Measures to keep pollution loads to a minimum such as the use of a minimum of detergent and alkali are possible but there may be a pay-off between consistent quality of cloth and chemical usage.

The average throughput of cloth at the time of our visit was 120 pieces per week and this made the effluent charge 6.5p per piece or roughly 0.15 per kilo of cloth processed.

Chapter IV

Dyeing and Finishing of Synthetics

INTRODUCTION

The Man-made fibres industry has probably the least end-of-line pollution problems, but the rapidly changing technology of process and product means that several in-plant measures have been taken. The nine cases presented here give an excellent coverage of the industry as they include independent and integrated concerns.

(i) PRODUCTION PROCESSES AND CONTROLS

There are no natural impurities that must be removed from synthetic fibres as in the case of wool, but treatment is needed for the removal of sizing agents, antistatic agents and lubricating oils used in weaving. There are different sizing requirements for spun and continuous filament yarns and the latter require less size of different types to the former. The finishing processes are similar for all fabric types (acetate, acrylic, nylon, polyester and rayon) and include scouring, initial rinsing, dyeing or bleaching, second rinsing and final finishing. The effluents are generally more dilute than those from cotton or wool finishing processes. Table 4.1 shows the range of BOD loads (expressed in pounds) per 1000 lbs of cloth processed for synthetic materials; we then give a more detailed account for each material type.

TABLE 4.1 RANGE OF BOD LOADINGS PER 1,000 LBS. CLOTH

Source	BOD (lb)
Natural impurities	3 - 7
Sizes, oils and antistatics	5 - 60
Scouring	5 - 50
Dyes, emulsifiers, carriers etc.	2 - 300
Special finishes	2 - 80
TOTAL	17 - 500

Acetate

Three stages may be used:

(a) Scour and dye. This is performed in the same bath and instead of scouring followed by dyeing, there may be scouring followed by bleaching. Sizes may be solubilized by diastatic or proteolytic enzymes prior to scouring. Dyeing is carried out using dispersed dyes, dispersed developed dyes, acid or napthol dyes. Bleaching is usually carried out with a 3 percent solution of hydrogen peroxide. Sources of BOD load are antistatics, sulphonated oils (used for lubrication) detergent and softeners. The typical BOD load is 2,000 ppm at 50 lbs per 1,000 lbs of fabric.

(b) Two rinses. These yield a waste with a low BOD loading, the source of which is carryover from the scouring and dyeing operations.

(c) Speciality finishing (where required). We have no information on "typical" BOD loads here.

The nature of the equalised waste from these three stages is about 9,000 gallons of water per 1,000 lbs of fabric, at pH 7-10 with a BOD load between 250 and 700 ppm at 20-60 lbs.

Acrylic (e.g. "Courtelle")

The following stages are used:

(a) first scour, followed by a rinse
(b) first dye, followed by a rinse
(c) second dye
(d) second scour, followed by a rinse, and
(e) speciality finishing

These types of fabric respond only to light shade dyeing unless pressure dyeing is used or unless carriers are added to the dye solution. For example, a cationic dyeing assistant will be added to acid dyes. Between 0.02 and 0.1 lb of carriers (such as phenols) will be needed for each pound of fibre dyed. The characteristics of the equalized waste from these stages will be about 575 ppm BOD, at pH 1.5-3.0 containing 25-50 ppm copper and 100-150 ppm phenols; as a consequence of the latter the waste may include toxic or taste producing substances. Carriers are not in general use with acrylics. The cuprous ion method of dyeing is no longer used.

Nylon

The following stages may be used:

(a) scour for removal of process chemicals (BOD load 1,350 ppm)
(b) rinse - low BOD load from carryover only
(c) dyeing - BOD about 375 ppm
(d) rinse, and
(e) final finishing

About 15,000 gallons of water will be used for every 1,000 lbs of fabric and the equalized waste will have a pH value between 7 and 9.5 and a BOD of 350 ppm. If waterproofing is carried out, a substantially higher BOD load will result.

Polyester (e.g. "Terylene","Lirelle")

The following stages are used:

(a) scour using non-ionic synthetic detergent
(b) rinse
(c) dye using high temperatures or a carrier such as phenylphenol or biphenyl if necessary
(d) scour
(e) rinse, and
(f) speciality finishing

This type of fabric has the highest pollution potential of all synthetics and the principal source of BOD load is the dyeing assistant used; if an effective substitute could be found, the pollution load could be reduced by half. There is also a significant contribution to the BOD load from antistatic agents. Characteristics of the equalized waste (about 15,000 gallons per 1,000 lbs of fabric) are a BOD up to 800 ppm. Polyester is almost always dyed without carriers.

Rayon

Two stages may be used:

(a) scouring and dyeing in the same bath, and
(b) a salt take-off in a separate bath (this is a rinsing stage with salt added to the water). About 5,000 gallons of water are needed for every 1,000 lbs of fabric and this is likely to have a BOD of 1,400 ppm, a salt content of 2000-6000 ppm and a pH value between 7 and 9.5

In addition, speciality finishing may be performed and this will lead to a substantial increase in the BOD load as a result of wax, solvent, gelatin and acetic acid used in the bath.

(ii) CASE STUDIES

Case 6

This plant, which is part of a large integrated group, is most unusual in that the dyehouse is part of a large petrochemical complex so that although the dyehouse is one of the largest in the U.K., it does not contribute any significant problems to the overall consideration of the combined site effluent. Further, the integrated nature of the plant has important implications for water and power considerations in the dyehouse. For example, water is extracted from a river and purified for use in the whole plant and similarly the combined effluent streams are pretreated before discharge to the sewerage; both these processes are subject to substantial economies of scale. In addition, steam is not generated on site but is supplied by a nearby power station for which the plant may be considered to act as one large heat exchanger.

Water and effluent treatment for the whole plant

A total of about 400,000m³ of water per day is extracted from the river by the parent complex; most of this is directly used for cooling purposes and is returned to the river but 36,000m³ are used for process water and of this, about 7,000m³ are used in the dye and printworks. The water is treated in one of two ways: 'filtered water' is produced by treatment with alumino-ferric and lime, with subsequent filtration through sand filters. 'Softened water' is produced by pre-treatment with soda-ash and sodium aluminate, with pH adjustment after settling and filtration, and subsequent base exchange. Most of the water used in the dyehouse is softened water at 'zero hardness' and this is used because a consistent water quality is needed and the river water has been found to be of variable hardness.

The process water used on site (35,000m³) ends up as a trade effluent which, after pre-treatment on site, is sent to the local sewage works. Under a legal agreement between the firm and the local corporation (now transferred to the water authority), the firm contributed one-third of the capital cost of the sewage works and pays an effluent charge; the consent conditions cannot be tightened before 1984. Part of the reason for this agreement was that the combination of normal sewage and effluent from this firm can be treated more economically than if they were separate - for reasons of their chemical consistency as well as the normal economies of scale. Details of the consent conditions are as follows:

pH	between 6 and 9
Suspended solids	160 ppm
Five day BOD	750 ppm
Oxygen absorption	110 ppm
Weekly BOD load	less than 162,000 kg.
Weekly max.volume	250,000m³
Daily max.volume	50,000m³
Hourly max.volume	2,700m³
Maximum temperature	38°C

In addition to these requirements, there are also restrictions on the daily averages taken over periods of one week and three months for suspended solids, five-day BOD and oxygen absorption. Further, the effluent should not contain more than 2 ppm of arsenic, chromium, copper, lead, zinc or cyanide (individually or in total)nor should it contain any grease, oil, tar, yeast or sugar and should not contain more than 20 ppm of sulphides.

A certain amount of pre-treatment is required to bring the combined effluent streams from the various parts of the plant to a standard within these consent conditions. All effluent is brought into one single stream and as this is usually acidic, lime is added automatically to neutralize it (the amount of lime needed is determined by a pH meter); the effluent then flows through open culverts to two settling tanks. The latter are large "Dorr-Oliver" concrete tanks; the first is a mixing section and the second a settling section. A travelling crane continuously sweeps sludge to the far end of the second tank where it is taken out at intervals and put into lagoons. This sludge contains mainly calcium sulphate from the neutralization but it also contains some organics such as cellulose derivatives; it is taken to drying beds periodically and then tipped. Liquor from the lagoons is recirculated to the front of the treatment tank and the fully pre-treated effluent is pumped from the settling tank to the local sewage works.

Annual costs of effluent treatment for the whole site totalled £300,000 in 1976 and of this £120,000 was the charge from the local sewage works and the remainder was made up of fixed and variable costs of on-site treatment. Fixed assets associated with effluent treatment were estimated at £210,000 and the average depreciation time was taken to be 16 years.

The dyeworks

The dyehouse buys in knitted fabrics from other companies within the group and sells finished goods to a variety of customers including large chain stores. It employs 400 people and throughput is 6,000 pieces (75 tonnes) per week; half of this is nylon 6, a quarter is acetate and the remainder is polyester and various mixtures. Although the actual size of the dyehouse is quite small, the level of technical expertise and added value are high; much of the output has to be treated to very high standards, as will be seen this can lead to additional pollution problems.

There is no need for elaborate scouring processes prior to dyeing since no sizes are added to knitted fabrics in the way that they are added to woven fabrics. The greatest amount of lubricant that it is found necessary to remove arises from polyesters and this is in the range of two to three percent; these lubricants are removed in the dye bath since they are self-emulsifying in water. In addition to dyeing, there is a small associated printing shop but this presents no serious pollution problems.

The charge for effluent treatment for the dyeworks and printworks was £80,000 in 1976 and the charge is calculated roughly in proportion to water consumption. This represents a higher unit cost than that paid by other firms in the group in the same type of business but it would not be in the firm's interest to treat the dyehouse effluent separately; in fact, this would probably increase overall costs since the dyehouse effluent tends to dilute the other effluent streams, thereby *reducing* the amount of pre-treatment needed by the other effluent streams. Since the effluent is treated centrally at a fixed cost, there is very little incentive to reduce pollution *loads* but there is more incentive to reduce water consumption. The firm is currently investigating the use of different dyeing machines with lower liquor ratios and with the group has conducted some research into solvent dyeing. One particular approach has now been wound up since the solvent does not swell the fibres as does water and hence the fabric does not achieve the desired quality; a further difficulty was that all available dyes were soluble only in water and suitable dyes were not available for solvent dyeing.

The total raw water bill charged to the dyehouse and printworks is of the order of one third of the cost a firm using town water would have to meet. In fact, the firm charges a lower rate for cooling water than for process water in accordance with the different amounts of processing each type requires. The water cost in the dyehouse is three percent of added value and effluent treatment costs are a similar proportion. Value added in the dyeing process (that is, excluding the cost of fabric) was about 89p per kilo of fabric treated; effluent treatment costs were 2.67p per kilo and the total effluent and water cost about 5.5p per kilo. However, fuel is by far the largest cost in the dyehouse; a typical monthly bill for steam, electricity and gas was about seven times that for water. These differences would be even more acute if it were not for the fact that this firm is able to obtain its steam very cheaply from the power station nearby - others would not have this advantage. Nevertheless, there is an on-going incentive to reduce the fuel cost and introduce better heat recovery schemes; about £50,000 p.a. is being spent in the dyehouse on heat recovery systems and this is expected to amortise itself in three or four years. The payback period for any insulation is one year. As yet, no heat is recovered from the drying process although it has been estimated that savings of up to one quarter may be possible; the most important factor in economical drying is to squeeze out as much water as possible in advance and in this respect synthetics are better than natural fibres since the latter may retain as much as 50 percent water by comparison with 10-20 percent retention for synthetics.

The firm experiences problems with stenter fumes and, in fact, considers that this will be its main area of pollution control expenditure in the near future. Elsewhere these stenter fumes are not encountered since setting takes place *after* dyeing but the high standards required by the trade in the U.K. mean that setting must take place before dyeing when, or course, lubricants still adhering to the fabric are burnt up by the setting process. A lowering of the standards set by customers would be almost imperceptible by the public and could lead to the removal of this source of air pollution, but possibly at the expense of a slight worsening of the water pollution problem since then the oils would come out in the dyeing process. An expenditure of £60,000 was anticipated on plant to overcome these fumes by burning them and recovering some heat although it was anticipated that the plant would not work continuously; a similar plant at another dyehouse in the group works on average on only 25 days per year. As an alternative, the fumes could be condensed but this would create difficulties with the disposal of a very viscous material.

<u>This case is unique and it is doubtful whether the similar situation that led to its establishment will arise again. There are advantages for an industry if the waste products of one (like low-pressure steam or large volumes of relatively unpolluted water) can be of use to another</u>.

<u>Case 7</u>

This firm, which employs about 190 people, is a commission dyer dealing mainly with synthetics but also with small amounts of cotton and some wool; both garments and fabric are dyed and the proportion of fabric dyed has been increased. This has had a beneficial effect on pollution since the dyeing of fabric needs substantially lower liquor ratios than the dyeing of garments - 10:1 (which is still relatively low by comparison with the rest of the trade) and 20:1 respectively. Weekly throughput of fabric is about 6½ tonnes scoured and 9½ tonnes dyed and throughput of garments about 5¾ tonnes scoured and 4¾ tonnes dyed; the total weight of material dyed is therefore about 14 tonnes per week. In addition to this, small amounts of yarn and hose are also dyed.

Total water usage is about 160,000m³ per year (3,200m³ per week) and this is either obtained from a spring (which is licenced for an extraction rate of 225,000m³ per year) or from the river (which is licenced for an extraction rate of 45,000m³ per year); a further 15,000m³ of town water are used for the boilers and this costs 9.57p/m³. Between 1975 and 1976 the effluent charge doubled to 3.30p/m³, and further increases were expected thereafter; the annual effluent charge was approximately £5000 in 1976. All water is softened since this not only protects the boilers but it also reduces fuel costs but the cost of softening is very high, about 6.39p/m³. This is composed as follows:-

Chemicals	1.76
electricity	0.99
labour	2.62
interest and depn.	0.97
maintenance	0.07

Any wool processed may contain up to 15 percent by weight of oils whereas the synthetics will contain only 5 percent or less; wherever possible, combined scouring and dyeing is carried out in order to save water and energy costs and this has necessitated the installation of specialised equipment. One high temperature jet dyeing machine was installed in 1967 and today similar equipment would cost £50,000; recently £18,000 had been paid for three high temperature garment dyeing machines.

All effluent is taken directly to the trade sewer without any form of pre-treatment; no balancing or heat recovery is practiced by the firm although it is considering the latter. All bleaching machines, which give off chlorine dioxide fumes, are connected to a stack to vent the fumes away but the firm was unable to estimate how much this cost - a figure of £5,000 would have been considered on the high side.

We estimate that the cost of obtaining and softening water was about 0.73 percent of total costs and the effluent charges 0.51 percent of total costs. These figures may be compared with the cost of energy which accounted for between 12.5 percent and 15 percent of total costs.

This seems to be one of the few firms that are relatively unconcerned about effluent costs and consent conditions. There seems to be an almost complete acceptance of the necessity of having to pay up when charges rise and a tightening of the consent conditions seems to raise few problems to management's thinking.

Case 8

This is a firm of independent commission dyers dealing largely with lace cloth which is now mainly made from synthetic material. There are approximately 90 employees, working a day shift five days per week together with a night shift on the drying side only.

Throughput of cloth is about 480,000 metres per month but the firm told us that it was difficult to translate this into a weight since the cloth is of variable width.

Water consumption is about 900m^3 per week (45,000m^3 per year) and until recently, this was extracted from a borehole but when the solids content became too high, the softening plant was unable to cope and the firm has now switched to using town water.

This cost 6p/m^3 galls. in 1975, (annual cost £2,500) excluding any costs incurred for softening and the firm saw this as more important than the 2.2p/m^3 charged for effluent disposal, even though the latter had increased substantially as a result of the water authorities' policy of equalisation and was expected to increase substantially again before 1980. Effluent charges were £1,000 p.a. but the firm considers the most important problem to be that of energy saving although, of course, reducing water usage is one way of reducing energy consumption.

The normal processing sequence is: scouring (to remove knitting oils) setting, dyeing, drying and inspection. However, about one-seventh of the throughput is 'rigid' lace which, in order to meet customer specifications, must be set by setting before scouring (if this is not done shrinkage is excessive); after this the lace is scoured and dyed in the same bath and then dried. Solvent scouring is used for about half of the throughput and the firm now has one continuous and one batch solvent scouring machine; the solvent perchlorethylene, is recovered although about 5 percent is lost. The residue is removed by a tar distiller who makes no charge but does not pay for what he removes. This change to solvent scouring reduced water consumption by one third and is one factor in its favour; further, it costs no more than conventional scouring with detergent and has the added advantage that since the cloth is not wetted before setting, the time taken is reduced by two-thirds with consequent energy and time savings. Thus the firm would like to process all its material in this way but cannot do so because, as a commission dyer, it has no control over the lubricants used on the fabrics reaching it and some lubricants contain graphite which is not suitable for removal by solvents.

Liquor ratios in dyeing are about 30:1, all dyestuffs were said by the firm to be biodegradable and the use of dyes with toxic constituents is minimized - for instance, chrome dyes are only used when black is required. Very strongly acid or alkaline effluents may be segregated and dosed before discharge but there is no equalizing or other treatment in the plant. Although the firm is not considering pretreatment, it is shortly to look at water re-use and heat conservation. It seems that the biggest constraint on the firm is the lack of space since it is sited in the middle of a residential area with practically no room for expansion. The firm carries out its own monitoring of effluents only when it is experiencing difficulties - it usually relies on the water authority and a trade association.

Because the 'rigid' lace must be set before scouring, the knitting oils give rise to offensive fumes. The firm was intending to install an afterburner to arrest these fumes at a cost of £30,000; in spite of this large cost, there are potential savings on energy for stentering which should reduce fuel consumption by one quarter and hence pay for the arrestment equipment in two years. These savings arise since the heat from the combustion of the fumes is recycled to the stentering machine and no additional fuel should be needed after lighting up; the firm considers this the most economical way of dealing with these fumes.

We were not able to obtain further information on water and effluent costs or on value-added.

<u>This is a typical lace dyehouse situated in a built-up area of a town. The constraints of pollution coupled with those of space must limit expansion and add cost even without further tightening of pollution control limits.</u>

<u>Case 9</u>

This dyehouse is part of a large vertically integrated group and is situated in a building which also houses a spinning mill for knitting yarn. The dyehouse is only a small part of the total plant operation; 1,650 people are employed altogether but only 40 of these work in the dyehouse. The latter works on a commission basis, dyeing small amounts of yarn to special order: the greatest bulk of the yarn is dyed during the spinning process when the dye is very easily absorbed into the fibre structure and dyeing costs can be cut to ¼ of the conventional costs. Therefore, at the time of our visit, with a slump in the trade, the dyehouse was working well below full capacity of 100 tonnes per week. The dyehouse is as fully automated as possible for this type of operation and recipes for dyes and chemicals are obtained via a computer.

Water usage varies directly with output and is approximately $100m^3$ per tonne; at full capacity $9,000m^3$ would be discharged every week. All water comes from the town main and is softened by passing it through resin beds which also serve to filter it. At a cost of $10.6p/m^3$ (1975 prices) the water cost amounted to nearly £11 per tonne of yarn processed; in addition, there was the cost of softening (£2 per tonne) and effluent disposal (£10 per tonne) so the total water cost was £23 per tonne. This represented about 4 percent of total operating costs although it excluded some small items such as the cost of a balancing tank and the in-plant changes which have reduced water consumption. In the period 1974/75, effluent charges were $7.66p/m^3$ compared with 4.19p in the previous year; both this and the water charge were expected to rise by about 15 percent per year after 1976. The firm had only been paying for its effluent for a short while since, when it initially set up in business, it was encouraged to do so by the local authority (which was concerned about area employment) with the inducement of zero effluent charges: such considerations being foregone with the 1973 Water Act which required the ending of preferential and discriminatory charges.

The yarn does not require pre-treatment before dyeing and since the dye uptake is about ninety-seven percent, the effluent is relatively weak. The characteristics are:- pH 4.5, BOD 120-500 ppm and suspended solids, nil. The use of cationic retarders - necessary to achieve even dyeing - increase the BOD load of the effluent significantly. The acidic pH value was outside the existing consent conditions (range 6 - 9) but until a short while before our visit the firm had been operating under an agreement with the local authority rather than consent conditions from the water authority and the former was prepared to accept the acidic effluent since it helped neutralize the alkaline domestic sewage. This agreement was not expected to continue in the future and the firm anticipated having to meet the consent conditions as from April 1977. (This may increase treatment costs at the sewage works and entail the firm in an expenditure of £5,000 on automatic dosing plant.) Depreciation and the cost of caustic soda will work out at about 1.10p/m^3 and add about five percent to water and effluent costs.

The most important in-plant change having an impact on pollution volumes had been a change from hank to muff dyeing; a muff is a package of yarn wound loosely so that it retains its roundness and in this way about twice as much yarn can be fitted into a dyeing machine than when conventional hanks are used. Thus water usage fell from 225m^3/tonne to 100m^3/tonne and the amount of steam needed also fell proportionately; there are also other advantages to muff dyeing since it achieves a more even dye quality and labour time is reduced substantially. Although more than three-quarters of dyeing is carried out in this way, the remainder is not suited to the technique and has to be dyed in hanks. The dyeing machines were easily modified to take the muffs at a total cost of £80,000 - £400 to convert each machine and £6,000 each for new stainless steel carriers for the muffs. The dyeing machines operate at atmospheric pressure and up to 110°C and a high volume of liquor is passed through with little force; the dyeing time is the same whether hanks or muffs are used.

The firm was planning a heat recovery system and also a full water reuse system but these were to be delayed until trade improved. At the time of our visit, dye liquor was heated via steam coils in the machines and after dyeing, cooling water passed through the machines and this was subsequently used as process water; the hot dye liquor went directly to the sewer. A more efficient way of utilizing heat would be to displace hot water over the top of the machines by means of cold water entering through the bottom of the machine (a fairly distinct boundary between the hot and cold water can be achieved) and by passing the displaced hot process water to a tank and then to a heat-exchanger. It was estimated that this would save about half of the existing use of steam in the dyehouse and at a throughput of 50 tonnes per week, the capital cost of £100,000 could be amortised in two years. Further, by means of a secret process it would be possible to recycle water about twenty times giving a reduction in water usage of about 80 percent and a saving of £7,000 p.a. at a throughput of 50 tonnes per week, for an initial cost of £100,000.

The situation of this dyehouse setting up in an area of high unemployment with the inducement of the prospect of no effluent charges is not untypical; successive authorities have altered some agreements although legal obligations are continued. On the technical side, the change of dyeing package to give a much reduced liquor to goods ratio is worth noting as a general principal in water saving.

Case 10

This dyehouse, which employs 250 people, is part of a large group although it operates entirely on a commission basis dyeing fabric for the fashion trade. As it is geared to this trade, the types of fabric processed vary considerably - for instance, there is now more woven and less knitted fabric and more cotton than a few years ago. Throughput is largely polyester, with some cotton and some viscose rayon in various blends. Pollution loads are dependent on the types of fabric processed but are not high; a recent trend towards fabric from spun yarn rather than filament yarn has increased BOD and COD loads, possibly because larger amounts of size are used in the former than in the latter.

Process water is taken from the river and is filtered and softened; boiler water is taken from the mains supply and this is chemically dosed before use. Between 9 and 11,000m³ of water are used every week. The firm used to discharge direct to the river but now no discharges of any type are allowed into the river and money has been spent to prevent leakages of even quite innocuous water to the river; alternative means of dealing with back-washings from the filtering system and with salt liquor arising from regeneration of the ion-exchange column (both of which used to be discharged to the river) have had to be found. Now effluent is discharged direct to the sewerage and the present charge for this is £2,000 but it is expected to rise to £40,000 in two years time following charges rationalisation. This increase in effluent charges is sufficient to put the future of this type of business at risk. The effluent constitutes one-sixth of the total effluent flow of the local sewage works which is experiencing difficulty in meeting the required standards at its present capacity.

Polyesters used to be dyed at atmospheric pressure and consequently the use of carriers such as o-phenylphenol was essential. Now, with a change to high-temperature dyeing use of carriers has been reduced by about 90 percent giving a cleaner effluent and also significantly reducing water consumption, albeit at the cost of greatly increased energy consumption.

WIRA have indicated that they consider there is no point in the firm undertaking its own effluent treatment although a limited amount of screening is done at present. The firm has looked at various ways of equalizing its effluent but this would not help to any extent. Water reuse and heat recovery are being investigated but, given the economic constraints, it is unlikely that anything will be done. We calculate that, at present, effluent charges amount to about 4 percent of production costs.

<u>This firm has a peculiarly local difficulty with a hard-pressed Local Authority. It would seem that a compromise is needed to spread over a longer period of time the inevitable high costs both will have to meet.</u>

<u>A good case for local cooperation in the best interests both of ratepayers, workforce and the Company.</u>

Case 11

This firm employs a total of about 400 people, of whom 140 work in the dyehouse. Throughput is almost entirely warp knitted nylon plus a small proportion of polyester and some cotton. The dyehouse is the third largest of its type in the U.K. and handles about one-tenth of the total volume of U.K. warp-knit nylon. The plant was entirely re-equipped in 1968 and since then substantial increases in capital productivity have been achieved and there is on-line computer control of machines with dye-mixes being prepared automatically.

Water, which is used at a rate of 1100m³ per day, is obtained from a borehole and hence its cost is very low; from a price of 0.05p/m³ in 1975, the cost rose to 0.09p/m³ in 1976. Nevertheless, the firm anticipates that costs of extraction will continue to rise in the future although it is unlikely they will reach the level of that of town water. The water is softened by removing bicarbonates and then it is passed through a base ion exchange column to complete the treatment and achieve the desired pH. The water is also heated by exchange with hot effluent.

The firm does not carry out any scouring processes since the fabric is knitted (and not woven) and there are no significant quantities of lubricants or other matter to remove. Dyeing is carried out in horizontal high pressure machines. Effluent is discharged into the trade sewer; the cold effluent goes direct but hot effluent is first diverted to a storage tank and then to the heat exchanger. The temperature of the hot effluent is about 75°C and this normally drops to 30°C, so more effective heat recovery could operate. If the hot tank is full, further hot water overflows into the cold water tank; the flow of effluent down the sewer is controlled by means of a valve to ensure an even rate of discharge. Discharge costs in 1975 were 1.38p/m³ (£4,000 p.a.) and in the year 1976/77 they were expected to rise to 5.26 p/m³ (£16,000 p.a. for the same volume of effluent). The firm considered that any "in-plant" measures to reduce pollution load and effluent volume - for example, the installation of a recycling system - would be very expensive. To a large extent the automatic controls integral with the plant have an effect on pollution loads; for example, chemicals and dyes are measured out accurately and in this way little is wasted. Pollution loads and effluent volumes are directly proportional to throughput.

Although at the time of our visit the firm did not treat its effluent in any way before discharge, the firm had considered the possibility of installing filters to reduce the solids loading of the effluent and hence reduce the pollution load. A full treatment plant had also been considered but both schemes had not been considered worthwhile at the time. However, one measure that was likely to be undertaken was the sale of some effluent (about one-fifth of the total) to a neighbouring firm as washing water; that firm would pay a small sum for this and also pay for all necessary pipes and pumps. It had been estimated that this could reduce the dyehouse effluent charge by about £3,000 in the current financial year and by £8,000 in the financial year 1979/80. This scheme had been suggested to the firm by the local water authority.

The stenters had caused no fume problems but, although it was under pressure from no-one to do so, the firm was about to spend approximately £5,000 on increasing the height of its chimney to minimise the incidence of fumes.

<u>This is a typical example of a new Midlands dyehouse. It is an efficient unit and when increases in charges come along the firm will have few opportunities for off-setting them: this means that the impact will be particularly great since it is in a section of the industry under intense economic pressure</u>.

<u>Case 12</u>

This dyehouse, which employs 300 people, is part of a large firm engaged in the manufacture and marketing of hose. The capacity of the dyehouse is 180,000 dozen pairs per week which would amount to about 31,000 kg of material, but at the time of our visit the dyehouse was working at only 60 percent capacity. A very significant part of the dyehouse operation is the preparation of the hose prior to dyeing - they must be packed together in a particular way for efficiency reasons - and of the total labour force at the plant, only 35 work on dyehouse operations proper.

All water used comes from the town water supply and cost 11.39p/m³ at the time of our visit (by late 1977 it had risen to 12.05p/m³), weekly usage is about 1,800m³ and during July 1976 (a time of restricted use because of under-capacity working) the cost of water was just over £600. All water is softened by ion exchange although the firm recognises that this may not be strictly necessary. The firm does not keep detailed cost records for this item of plant but considers that operating costs are very low; a new softening plant had been installed recently at a capital cost of £5,000 so the annual capital charge is unlikely to exceed £1,000 and the main item of operating cost is the need to replace about 4 tons of salt every month in the regeneration plant.

Effluent is discharged direct to the trade sewer and thence to the nearby sewage works, no pre-treatment is carried out by the firm at all. Effluent disposal cost 10.79p/m³ and during the first six months of 1976 average monthly disposal was just under 6,000m³ at an approximate cost of £650 (£1,700 in late 1977). Total water supply, treatment and effluent charges worked out at around £16,500 in 1976 and the firm considered that this represented less than one percent of total production costs and therefore there had been little incentive for the firm to consider water and effluent problems. Until six years before our visit the firm paid no effluent charges at all and when the first charge of £1,500 was levied, there was some opposition to it. However, the firm told us that it considered that gradual increases in cost become assimilated very easily and that it would take very large and sudden increases for water usage and effluent disposal to become a major problem.

Water utilization did not feature in the choice of dyeing machines, rather labour utilization was seen as the most important factor; the firm designed its own dyeing machines which enable one third of production to be dyed utilizing only three operators at a time. Scouring and dyeing are carried out in the same bath and light shades need only two baths; darker shades may need up to six. There is therefore a significant difference in dyeing costs, depending on the colour to which the batch must be dyed. Further, the rate of water usage increases as production falls since the dyeing machines are not filled to capacity; the firm felt that future production would have to be rescheduled so that only large batches of the same colour were produced.

Because of the way in which the hose are packed into the dyeing machines, very large amounts of dyes and chemicals are necessary to achieve the penetration required. Thus, about one quarter of dyes and chemicals remain in the dye liquor and the effluent is strong - the COD values are of the order of 2,000 to 3,000 ppm. Disperse and Acid dyes are used and thus the pH of the effluent is slightly on the acidic side but the suspended solids content is very low. Recovery and reuse of dyestuffs from the effluent is probably not feasible but total recycling of water would be possible. In fact, this was seriously considered a few years ago when estimates presented to the company on the Mortensen system made the scheme appear very attractive. The capital cost would have been of the order of a quarter of a million pounds (at 1972 prices, perhaps double or treble today) and running costs were estimated to be about $5p/m^3$. This system was attractive especially since its space requirements were low and, in fact, the firm considered that there was no other system that would be suitable for its effluent. However, on detailed examination it was found that running costs had been seriously underestimated and the scheme was abandoned. In the long term such a scheme would prove beneficial but the firm felt that board approval would be unlikely unless water and effluent costs rose dramatically. As we have already said, there is no treatment of effluent at present although a cooling tank is used to lower the water temperature to $43^{o}C$ before discharge.

In spite of the *financial* incentive to save water, the recent drought had acted as a catalyst to water conservation; usage was reduced by 20 percent by means of the reuse of selected finishing liquors from the dyeing process. The firm felt that, if necessary, more could be done to reduce water consumption.

<u>This company voices an opinion which is widely held. That is the small effect on costs of water and on margins when it is possible to absorb small increases gradually. In the short term this makes decisions to spend large capital sums on recycling difficult to justify. It does not alter the pressures which will eventually be brought to bear on effluent quality which will need expensive projects on site, i.e. before discharge to the local authority sewers.</u>

<u>Case 13</u>

This firm acts as commission dyers of polyester and cotton fabrics and also dyes some garments and hose. There are 400 employees and at the time of our visit the throughput of fabric was 45,000 kg per week although it was expected to rise to 60,000 kg. About the same weight of garments and hose are dyed.

Total water usage is currently about $18,000m^3$ per week and this comes from two possible sources; a stream and a canal. The firm holds a licence to extract $771,800m^3$ annually at a cost (in 1975) of 2.20p/m³ and a licence to abstract 454,000m³ annually from the canal at a cost of 8.48p/m³. When we visited, the stream had run dry due to the drought and therefore the firm had to abstract from the canal; and even in 1977, the amount of water available from the brook was very small so that most water had to be drawn from the canal. The estimated annual bill for water was therefore of the order of £30,000. Further, the water is softened in a Permutit plant and capital and operating costs for this must be included; the plant cost £150,000 to install in 1969 and a realistic capital charge would therefore be at least £30,000 per annum and on top of this should be added an unknown amount for current operating costs on this plant.

Effluent is discharged to the sewerage without pretreatment (although some heat recovery is practiced in parts). The limits set on the effluent are COD 1,200 ppm., suspended solids 400 ppm, and pH range 6 - 11; in addition, there is a limit on BOD values which is currently being exceeded. Effluent charges for the financial year were expected to be about £24,000 but since the local water authority had not completed the phasing in of increased charges it was felt that charges for the same volume of effluent could be £50,000. The total cost of water and effluent disposal could therefore eventually exceed a six-figure sum.

The firm had had under consideration a total recycle system and manufacturers quotations suggested that this would cost £250,000. However, there could be no guarantee of the purity of the recycled water and it would not always have been suitable for use in processing and it had therefore been decided to abandon the scheme. Instead, the firm was to negotiate higher discharge limits with the water authority, increasing permitted discharge from 3,300m³ to 3,600m³ per day. After a two year period, the firm will pay a capital charge for increasing the capacity of the neighbouring sewage works; this will involve the installation of plastic media towers to reduce BOD loads and, in fact, a pilot plant was to be installed shortly after our visit. The firm hoped that eventually there will be sufficient capacity to cope with 4,500 m³ of effluent per day. In addition to the extra equipment at the sewage works, the firm will have to install equipment for equalizing the effluent and it was estimated that this would cost £60,000.

The precise nature of the effluent is determined by the specific mix of materials to be dyed; there has been a trend to more cotton goods and this had resulted in more pollution and higher costs. More labour and more water is required and Reactive rather than Disperse dyes were used (the latter cost only £4/kg compared with £7/kg for Reactive dyes); Reactive dyes also give rise to higher BOD loadings in the effluent. Further, cotton must be pre-scoured to remove sizes (which are commonly based on paraffin wax) and bleached with peroxide; both these processes increase effluent strength. As the trend is to cold dye cotton, the polyesters may be dyed in the jet-dyeing machines and by the time we visited all polyesters were dyed in this way without the use of carriers. Wherever possible, scouring and dyeing are carried out in the same bath. All cooling water is recycled and some heat is recovered from hot liquors from the jet machines.

The firm estimated that steam accounted for 23 percent of sales value and water and effluent together accounted for 6 percent of sales value.

This is a company which has had to spend large sums on water in all its aspects. The decision against recycling shows that much R & D work needs yet to be done in this area. The advantage of a good relationship with the local authority also comes through.

Case 14

This large dyehouse, which employs over 500 people, acts as a commission and in-house dyer for a large group. It dyes a wide variety of materials but the present throughput of approximately 170 tonnes per week is divided roughly equally between artificial and synthetic fabrics. Since practically all fabric comes from within the firm, the dyehouse has some control over the lubricants or sizes used in the manufacture of the fabric but this control is very limited and usually the dye house is able only to ascertain which sizes have been used so that it may use appropriate dyes.

The plant is located close to a river but all water is extracted under licence from a borehole; since this water is very hard and dyeing would be impossible without it, it all must be softened by ion exchange before use. The cost of abstraction and softening (including capital charges) is such that the cost of this water is about one-fifth of what the cost of town water would be and we therefore estimate that the cost of water supply is about 2.2p/m³; as about 24,000m³ are used every week, the annual water bill is approximately £25,000. Effluent is discharged after treatment to the river which for a long time has formed a source of potable water lower downstream; because of this set-up, the firm is effectively supplying the water authority with water and it has been suggested to the firm that the installation of a full re-cycling system would not be in the water authority's interest! Because of its distance from the works, the firm has no opportunity to use the local sewage works and even since before the last war the firm has been fully treating its effluent although of course, the standards required of it have been gradually tightened from an already high performance. At present the consent conditions are:- five day BOD 15 ppm maximum, suspended solids 20 ppm maximum, and pH 6-9: in addition, there is an "unofficial" chloride limit of 500 ppm which arises because the waste water is eventually used for irrigation purposes. These standards are tighter than the Royal Commission recommendations but the firm is able to meet them although there are currently negotiations in process over possible future specifications for the colour of the effluent. Although the effluent is treated to a very high standard, it is unsuitable for reuse in the plant and to do this would require additional treatment and even then some effluent would have to be disposed of which would require treatment to the current standards. The firm is presently experiencing some difficulties meeting the suspended solids requirement and envisages the installation of further treatment capacity for that in the future.

The principal contribution to BOD load arises in the dyeing process; final effluent characteristics from the production process were pH 7.5 - 8.3, temperature 18°C, BOD 150 - 300, suspended solids

50 - 150 ppm and total solids 1860 ppm. The high total solids content arises from the presence of water soluble detergents, dyes and sizes. The effluent is equalised after it has passed through two 220m³ receiving tanks in a balancing tank of over 2000m³ capacity; it then passes to an activated sludge plant which has a fairly high control efficiency of BOD, reducing this to about 30 ppm (a maximum control efficiency of 90 percent) but the efficiency for suspended solids is less, reducing this to a level of 120 ppm. Effluent leaves this plant a dark colour. It then passes to a chemical precipitation plant which further halves the BOD load and reduces the suspended solids content to a level of 70 - 80 ppm and leaves the effluent a pale colour. The effluent finally passes to a lagoon and a biological field which, whilst further reducing BOD load, has a greater impact upon suspended solids, reducing these to a level of 20 - 30 ppm. The final effluent is a pale colour and overall maximum control efficiencies achieved are about 95 percent for BOD and 85 percent for suspended solids. The liquid effluent treatment plant creates a ferric hydroxide slurry and about 45m³ per week are pumped to a nearby gravel pit where it is allowed to dry before it is dumped on the firm's own land.

The firm had not been able to lower the pollution load through the use of fewer process chemicals but efforts had been geared towards increasing the amount of fabric dyed in one batch and this had had the effect of slightly *increasing* the pollution *load* but reducing the volume of water used per unit of cloth processed. The firm had changed to biologically soft materials where possible and this had the effect of slightly raising the BOD load of the final effluent from the plant but reduced that at final discharge by about one-fifth and also reduced the incidence of foaming. In some instances this resulted in a product of slightly inferior quality. No materials are recovered from the effluent and although size recovery had been considered, it had been thought to be uneconomic, particularly since there is no separate scouring operation and size removal takes place at the same time and in the same machine as dyeing, and so a coloured size would be obtained. No form of heat recovery was in operation although the firm had contingency plans should the cost of energy rise substantially; the efficiency of the effluent treatment plant would be slightly impaired. The only incentive to reducing water consumption had been energy conservation and the firm had introduced no counter current or similar techniques.

The firm was aware of the problem of air pollutants and the use of ester based carriers for dyeing polyester and the use of stenters had led to one or two complaints. Chemicals were chosen with care and the fumes from the setting of synthetics were discharged through short stacks to the atmosphere.

Computation of the cost of control in this firm is complicated by the fact that some of the equipment is over 40 years old. A "conservative" estimate of the replacement cost of the control facilities is that it would exceed £500,000 and the firm anticipated that more needed to be spent in the near future on lagooning that would cost £25,000. The water saving measures had saved about £1,000 p.a. but the water quality control laboratory cost about £8,000 p.a. to run. The cost of effluent control was between one and two percent of turnover but, given the low margins in the industry, was perhaps as much as 30 percent of profit.

This dyehouse provides, perhaps, an example of what the industry may ultimately have to introduce if and when effluent standards are really tightened. Already the cost is a substantial burden and to replace the treatment plant at £½m would be serious. Fortunately, this firm is located in the country and has ample space for treatment plant. The firm has had an excellent relationship with the Local Authority over many years and this had led to realism, particularly since the area has little surface supply and falling underground resources.

Case 15

This printworks, which is part of a large, vertically integrated organisation, employs approximately 230 people and is engaged in the printing of a wide variety of spun and filament fabrics. Current throughput is approximately 20 tonnes per week. It is located on a comparatively new trading estate; the operators of which supply water to the works and also undertake effluent disposal. Approximately 10,000m^3 of water is used every week which requires only minimal treatment before it is suitable for processing. The water procurement cost is 8p/m^3 but effluent disposal is entirely without charge. The consent conditions for the discharge of effluent are not available but the effluent is regularly tested by the water authority. The company was unable to break down the individual sources of pollution loads but the overall effluent has a pH range of between 7 and 9, a COD load of up to 850 ppm, and a BOD load of up to 250 ppm.

As far as in-plant measures to reduce pollution are concerned, the company feels that it has been under no incentive to reduce pollution loads in the past 3 years by reducing the quantity of process chemicals used or by changing the types of chemicals used. No materials are recovered from the effluent and no heat recovery systems are installed. No specific counter-current or water re-use systems have been installed, apart from those that were already established in the process and equipment installed at the plant.

There are no end-of-line effluent treatment facilities of any sort; no effluent streams are segregated and there is no equalization of effluent before discharge to the sewerage. The only air pollution that arises is the very occasional discharge of lint to the atmosphere which is on a minor scale and for which the firm has no control device.

The only pollution control cost incurred by this medium-sized plant was the extra cost of water re-use systems incorporated in the plant when it was built. Situated in a non-textile area, this plant is probably not very typical of those located elsewhere for reason of the nature of its throughput and the non-existent effluent disposal costs.

Chapter V

Cotton Dyeing and Finishing

INTRODUCTION

The cotton industry has had substantial problems with both the quantity and quality of its effluent. Many changes of dyestuffs have taken place recently and these and other in-plant changes have certainly added to the industrys' costs - although there have been beneficial production effects as well. The cases here give a reasonable coverage of the types of firms in the industry and their pollution problems.

(i) PRODUCTION PROCESSES AND CONTROLS

The cotton finishing processes are summarised in figure 5.1. The first stage is desizing which must be performed before further processing and in this stage starch is hydrolysed by the action of enzymes or sulphuric acid. The retention time is between 3 and 12 hours and this is then followed by rinsing and then a run through a caustic and penetrant bath. Some of the organic matter in the waste arises from natural impurities and the other from the sizing material present in the fabric.

An alternative size such as carboxy methyl cellulose (CMC) can reduce BOD load significantly. However, the 20 day BOD load of this material is high so that by using CMC or mixtures of starch and CMC as the sizing agent one problem (initial demand at a biological treatment plant) may be solved only to be replaced by another. The desizing wastes are a potential source of glucose and by recovering this BOD reductions of up to 50 percent may be achieved. Recovery of natural compounds such as pectins and waxes *might* also be technically and economically feasible, but this is not widespread.

Dyeing is followed by scouring (sometimes referred to as kiering). The traditional (though now less common because of the use of cotton/polyester blends) method is the pressure kier boil in which the fabric is boiled for 2 - 12 hours at a temperature of 250°F. Process chemicals used are sodium hydroxide (1-8 percent of cloth weight), sodium carbonate (1-3 percent) and sodium silicate (½-1 percent). In addition pine oil soap may be used to remove

FIGURE 5.1

COTTON FINISHING WASTES

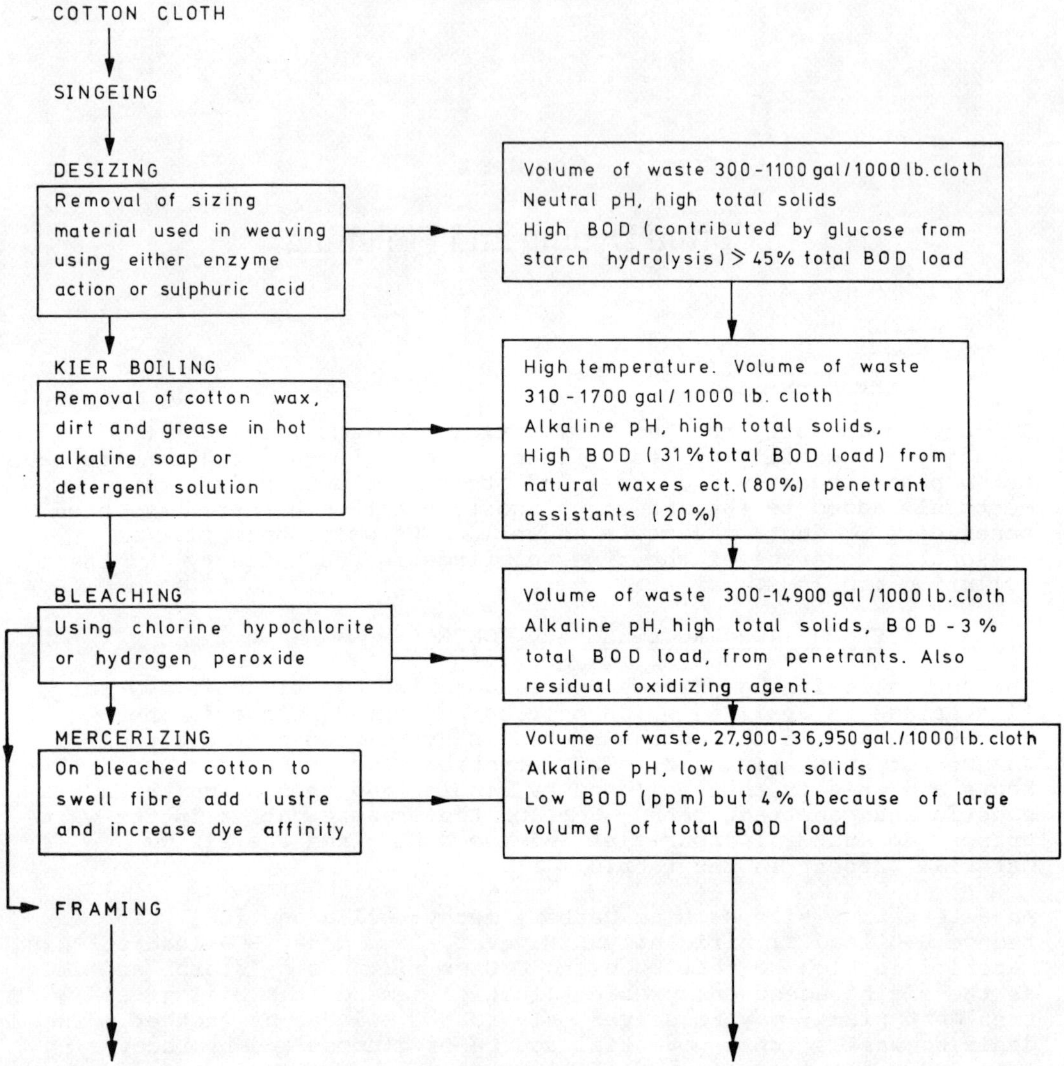

CONTINUED

FIGURE 5.1 contd

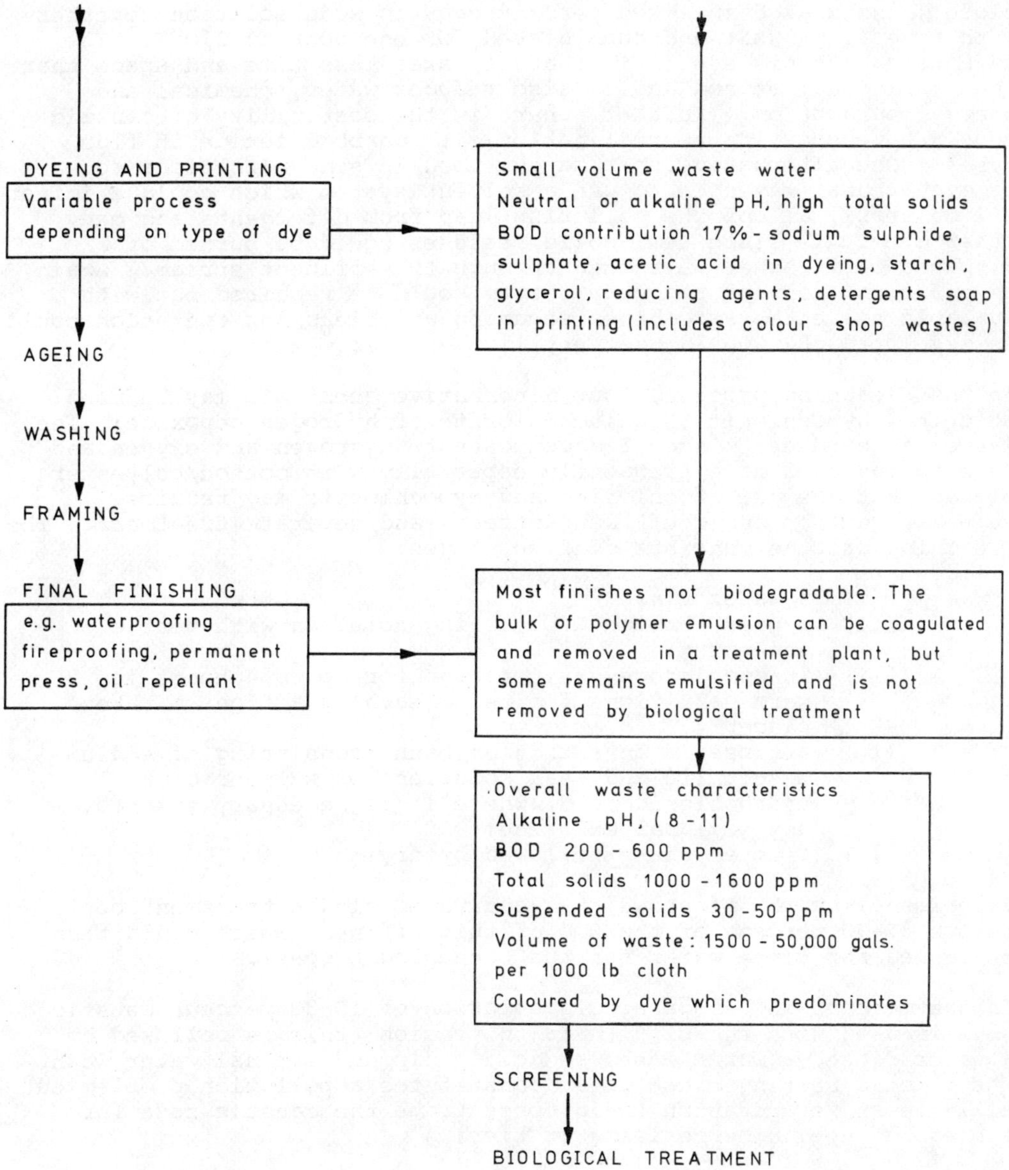

Adapted from "Pollution control in textile mills"
H. A. Schlesinger, Emil F. Dull, Thomas A. Fridy

natural waxes and fatty alcohol sulphates used to provide a melting action.

A newer method is the continuous scouring device in which the cloth is saturated in a 3-6 percent caustic soda solution together with a wetting agent and then heated for one hour at 210°F. The advantages of this system is that it takes less time and space than the traditional method and it also reduces water, chemical and energy consumption. Further, there is the possibility of caustic soda recovery or its neutralization with carbon dioxide in flue gases. One alternative that may be used in some situations to remove natural waxes is to use a solvent system which could eliminate all or nearly all of the pollution load from detergents and many other chemicals since semi-solid residues could be burned or disposed of in other ways than through the effluent stream. Again, an efficient solvent recovery system would be required but with suitable solvents, a combined scouring and bleaching operation could be developed that would use very little water.

In the bleaching process, four alternative chemicals may be used: chlorine, hypochlorite, sodium chlorite of hydrogen peroxide; The latter is harmless since it decomposes to hydrogen and oxygen and is becoming used more frequently especially with cotton/polyester blends, but the use of chlorine and hypochlorite may require segregation from other effluent streams and separate treatment. The bleaching process consists of five stages:

(i) a water rinse;
(ii) a pass through a bleaching solution with gentle squeezing;
(iii) in some processes, storage for up to 4 hours in warm conditions for the bleaching action to take place;
(iv) a rinse in an antichlor bath (consisting of sodium bisulphite and weak solutions of sulphuric or hydrochloric acid) where it is necessary to remove any residual chlorine;
(v) a water rinse followed by drying.

Experimental work has shown that activated sludge treatment could remove 85-90 percent of the BOD of this effluent which could then be reused for rinse water for the bleaching process.

In the mercerizing stage a cold solution of 10-30 percent caustic soda is used with the cloth held in tension; this is followed by a water rinse, a water wash, an acidic dip and a final water wash. The caustic bearing rinse water constitutes a pollution problem but dialysis and evaporation could concentrate the caustic soda for subsequent use in mercerizing or kiering.

The types of dyes used in the dyeing and printing stages may be classified according to their method of application:

VAT - a dye put on in a reduced state and oxidized; although difficult to apply, these dyes have excellent wet fastness and are widely used in shirtings, denim, "effect threads" and curtaining.

REACTIVE - a dye which reacts chemically with the cellulose molecules (i.e. the cotton itself). These are quite widely used.

SULPHUR - a dye put on in a reduced state and oxidized - again with excellent wet-fastness which makes it suitable for corduroy.

ANILINE BLACK - aniline oxidized by air or steam ageing.

DIRECT - dyes applied directly; there may sometimes be after-treatment to improve the wet-fastness and are usually only used for pale shades.

PRINTING - any of the above types of dyes applied by roller or screen print machine. Much of the wastes here arise from cleaning operations and excess paste on equipment.

WASTE TREATMENT

Wastes from cotton dyeing and finishing are generally treated by biological methods - chemical coagulation is generally not used because of the high content of soluble BOD. Nor is this method economical for chemical removal because a very large quantity of alum is required for full colour removal. However, when small quantities of strictly chemical wastes are present, coagulation may be suitable.

In the treatment of cotton finishing wastes, attention should be given to the following potential problems:

- the presence of toxic substances (such as mildew depressants or chromates from dyes) which may require segregation
- nutrient deficiency,
- pH range (it must be within 6.5-9.5 for biological treatment)
- the possibility of shock loads,

and

- the presence of synthetic detergents and waterborne waste fibres (fine screens are difficult to maintain and may have a short life). The best solution found to the latter problem is to select treatment processes which will tolerate the fibre and then settle as much as possible in a final clarifier and dilute the balance in a large volume of water.

Various treatment methods may be applied to these wastes and typical control efficiences for these have already been given in Table 2.1.

(ii) CASE STUDIES

Case 16

This firm, which employs about 400 people, is part of a large integrated group; it is one of the largest commission dyers and finishers in the group and in the country. The plant was virtually re-built with new equipment five years ago, at a capital cost of

£3 million. Throughput is about 110 tonnes per week of mostly woven cotton/blends although small amounts of synthetics are also processed.

Water usage is 25,000m³ per week; ninety percent of this is obtained from boreholes and brooks and the cost of this water is £85 per week for pumping plus an annual abstraction licence of £800. The remaining ten percent is town water for which a price of 11.67p/m³ is paid and naturally this is much more expensive than water from the other sources. The future supply of brook water is uncertain (not only because of the dry weather when we visited) and a greater recourse to town water would increase production costs significantly. Water is no longer pre-treated in any way; at one time it was softened but this was discontinued after it had been found that not carrying it out made no impact on dyeing quality. All heating of water takes place in the dyeing machines.

Discharge of effluent is at a rate of 230m³ per hour (the maximum permitted is 320m³ per hour) and the flow is controlled by means of a holding tank. There are also limits on pH, temperature, sulphides, oil, grease and various other specified parameters. Since the renegotiation of the consent conditions, limits are no longer set on COD loads or on suspended solids and this may be to the firm's advantage although, of course, the effluent charge increases as these factors rise. At the time of our visit, the effluent charge was £46,500 p.a. (i.e. 3.85p/m³) but this represented an increase of between three or four times the charges prevailing in 1975 and the firm was anticipating further increases in the future. Thus although the *conditions* were more favourable, the *price* movements had been strongly adverse.

The fabric has to be de-sized prior to dyeing and this generates the highest pollution loads, the effluent from this part of the process has a COD load of about 8000 ppm. Since not all the firm's input comes from within the group, the precise constituents of the size are not known with any certainty in advance and this means that the firm has little control over the de-sizing process. If the firm were totally vertically integrated, control of this process could be much more rigorous and it is possible that the size could be recovered. As it is, recovery is impossible because of the numerous different sizes used - starch, polyvinyl alcohol and carboxymethyl cellulose are common. The dyeing process itself produces a relatively clean effluent - the estimated COD load is about 100 ppm - but the volumes concerned are very high. Other operations carried out include bleaching, mercerizing, and various other specialised finishing operations. The firm has stenters but there had been no complaints over their operation and the firm did not envisage having to install any arrestment equipment for them.

The running costs of the holding tank and the associated pumps was estimated to be £1,400 p.a. and the replacement cost of the holding tank to be about £15,000 although the one used was old and fully depreciated. The firm estimated its cost of monitoring effluents at £100 p.a., to give a total identifiable pollution control cost of £48,000 p.a. which represented 2 percent of operating costs. Until the time of our visit the firm had not had to invest in any form of treatment plant for "end-of-line" control apart from the desizing of highly alkaline mercerizing effluent before it mixed with other effluent streams.

The firm had taken some very significant "in-plant" measures to reduce effluent volumes at source, but it was evident that the firm did not view these as pollution control but a means of making production more efficient. The new plant itself had substantially reduced water consumption; all the later dyeing machines had low liquor ratios and all but one of the scouring and rinsing machines had counter-current water flows. Thus in the late sixties before the new plant was installed, water consumption was approximately the same as at present but for a much lower fabric throughput. Further measures had also been taken so that achieved water consumption was about one-third less than that predicted at existing outputs. Conversion of one dyeing machine had reduced its water consumption by half and dye usage by ten percent and the capital cost of the changeover of £4,500 had been recovered in 18 months. The firm thought it likely that further similar conversions would be carried out. Another interesting - and highly cost-effective - way of reducing water consumption had been the installation of orifice plates on valves so that when valves were fully turned on water was not wasted.

Changes in the pollution load of the effluent were taking place continuously due to changes in types and quantities of chemicals used although pollution loads also varied considerably with the nature of the fabric processed. There was a continual effort to reduce quantities of chemicals and dyes used in order to reduce costs and, of course, this could have the effect of reducing pollution loads. Toxic chemicals and dyes were avoided and all chemicals were bio-degradable and non-herbicidal although the onus of ensuring the suitability of dyes rested with their supplier. Use of sodium sulphide - a reducing agent for some types of dyes - had been cut to a minimum and suppliers had been trying to find a suitable alternative. No heat was recovered from the effluent itself but some hot clean process water (for example from air compressors) was taken to a 115m³ tank and the heat from this water was recovered before its reuse.

Some research had been carried out into effluent treatment, starting at plant level and continuing at group level. In particular, since it was desizing effluent which gave rise to the major part of the effluent charges, segregation and separate treatment of this effluent had been considered. Flocculation with alum would have yielded a 40 percent reduction in COD load but the cost of alum alone, regardless of capital cost, would have made this uneconomic. Work had also continued with other, cheaper flocculants such as copperas (ferrous sulphate) which could yield COD reductions of fifty percent and which would have resulted in a saving of £8,000 p.a. (17 percent of the total) in effluent charges; we were unable to ascertain whether this could be justified on economic grounds. Treatment of the whole effluent stream had also been considered - aeration could reduce COD loadings by between 10 and 50 percent or a trickling filter by considerably more than this. Again, the firm was unable to give us any information relating to the likely costs of such plant.

<u>This is a dyehouse at which high costs are likely to rise further. Already very conscious of water costs, it has made significant</u>

savings by reducing the volume of consumption but eventually it will have to face the alternatives of increasing effluent charges paid to the Water Authority or massive in-house treatment.

Case 17

This dyehouse, which is part of a large group, employs about 250 people and dyes approximately 40,000 kg. of lining fabrics every week. A total of about 3,600m³ of water are used every week and this comes largely from a river and the remainder, about one fifth of the total, comes from the town supply. Given the very much lower cost of water from the river than from the town supply, we estimate that the total water bill for this company is of the order of £5,000 per annum, although this does not take account of the capital cost of the reservoir that has enabled the company to do this. All water is softened by a base exchange process and the firm calculates that this cost £88 per week or £4,500 per annum, although we do not know whether this includes a capital charge for the equipment used.

After treatment, all effluent is discharged into the river; the firm has spent substantial amounts of capital on its effluent treatment facilities in recent years. The characteristics of the combined dye and wash liquors is a pH of 8.4, a temperature of 22°C, a BOD load (5 days) of 1,130 ppm and a suspended solids load of 230 ppm. These loadings are clearly outside the prescribed ("Royal Commission") limits of 30 Mg/l for suspended solids and 20 Mg/l for BOD and the effluent must therefore be treated; it is discharged into a 340m³ balancing tank before treatment and then finally into large lagoons of approximately 4,500m³ capacity from which it percolates into the river at an average discharge 24 hours per day, 7 days per week. The treatment consists of screening, sedimentation, chemical precipitation (flocculation) and by means of a trickling filter. The characteristics of the treated effluent as it enters the river are a pH of 7.1, and ambient temperature, a BOD load of 12 ppm (implying a control efficiency of the order of 99 percent) and a suspended solids load of 13 ppm (implying a control efficiency of about 94 percent); these values are well within the prescribed limits. About 300 tons of solid waste are dredged from the lagoons annually and these are dried off in a sludge lagoon and subsequently deposited on the work's refuse tip.

The high capital cost of the treatment plant (which now accounts for about 5 percent of the capital value of all plant and machinery on the site) has been spread over a number of years and it is accordingly difficult to estimate the true cost of pollution control in the firm. In 1974, the operating costs for labour for the operation of the effluent plant, alone amounted to £4,750 and we would estimate that water procurement and disposal (excluding capital charges) currently amounts to about 3 percent of operating costs. This is possibly lower than other firms in the industry but is clearly a consequence of the measures taken by the firm to put itself in a position, not only to meet Royal Commission standards, on effluent, but also to recycle its own water. Naturally, the extent to which purified effluent is recycled is affected by the compatability with the raw water and the costs of electrical energy for pumping uphill to the storage reservoir.

The pollution load in this case arises from the size content of the fabric and not from the nature of the production processes themselves; accordingly the firm is not able to reduce pollution loads by reducing the quantities of process chemicals or by changing their nature. We have already mentioned above how the firm is now able to recycle water and in addition to this it now undertakes some heat recovery from the effluent.

This plant is an example of how the full range of on-site treatment facilities maybe used so that ultimately the firm is able to recycle its own water. This enables the firm to reduce its operating costs but imposes the penalty of very high levels of capital expenditure.
The figure of £4,740 for operating costs can be misleading. Although there is mention of the high capital costs there are no figures. One gets little for £250,000 over 20 years (a relatively long write-down) this is still £12,500 p.a., *more than double* the operating figure.

Case 18

This small company makes cloth tracing paper and hollands (the material used on the inside of tyres.) There are 25 employees, in a plant which was once part of a massive dyeing operation that shrunk in the face of Far East competition.

We believe the company to be a very good example of how a small firm can successfully innovate to overcome energy and pollution problems. It has also adopted a strategy of specialisation to avoid the large competitive legions.

The owner built his own condensate recycler from scrap parts at a cost of £8,000 which avoids the necessity of discharging effluent to the sewer and makes heat savings of £2,000/annum (1975 prices). Thus the firm hardly discharges any effluent at all, and is able to reuse waste starches.

This is the only example - apart from the wool processor in case one - where a firm actually recovers material from its effluent and is thereby able to reduce its overall costs of production.

Case 19

The firm is the largest manufacturer of corduroy in the U.K. and has a substantial share of the total market. It has only 3 competitors in this country and there is an absence of Far East competition; competition mainly comes from the EEC and United States.

The plant was set up 3 years ago by an American firm and we were told that the location decision was based mainly on free discharge of effluent, a sympathetic local authority, and cheap capital grants from central government. 50 tonnes of cord are produced each week on 4 continuous lines using sulphur dye-stuffs which whilst highly pollutant are costing *in this case* one third of vat dye costs for a non-continuous process. Water costs were 4.4p/m^3.

The firm has a highly alkaline effluent, which is sometimes discharged at a temperature of 60°C and contains sulphate at a rate of about 100 ppm in the form of sodium sulphate. The effluent is also highly coloured. Volume of effluent is 18,000m³ per week.

The firm appears to be a heavy polluter. Standards also vary substantially from what the council now want the firm to achieve although at present there is no pressure from them to do anything. Their requirements are exceeded by the firm's pH value, temperature, suspended solids and sulphate content. The firm has also caused a substantial nuisance to local residents from hydrogen sulphide which is emitted from the dye mixing room. Until recently nothing had been done but a new environmental health officer had forced the firm to begin to look for some type of treatment and had threatened to serve an order if the firm did not do this.

So far the firm had made *no* end-of-line pollution control expenditure apart from some small expense (£40 per week) to deodorise the airborne emissions from the dye mixing room. However they have now commissioned a new plant to deal with the whole range of treatment. Costs for this were not available.

There have, however, been in-plant changes. In particular the firm uses a non-metallic oxidising agent which costs £70,000 per annum more than one which would do the same job but which would cause heavy metals to be discharged into the effluent. Against this in-plant cost, the firm saves a great deal of money by using sulphur dyes which it claims are one third the cost of vat and ordinary dyes but which also have substantial advantages with the continuous dyeing process. The firm has by good housekeeping reduced pollution by using less water and ready mixed dyestuffs.

The major point of this case is that the firm is able to get away with massive pollution and has until now only made in-plant changes to abate pollution and with hindsight it seems that the local authority should have placed some limit - albeit a slack one - on the discharge from the new works. Current ecological thinking, however, is tougher on what were economic/political considerations in encouraging industry to the area.

Case 20

This firm is an old independent family concern employing 120 workers with a plant worth approximately £½million and with a throughput of approximately 13,000 kg. of cotton cloth per week. The firm owns a great deal of land and has its own reservoirs on it; the local sewage works is also close by. Water is gravity fed from the reservoir straight to the works without prior treatment and the firm's cost of water is £250 per year in abstraction licence fees.

Until quite recently effluent was discharged direct to the river with only acid neutralisation treatment and this effluent included some quantities of sodium sulphide. As a result of legislation curtailing pollution, more sophisticated treatment was necessary and at the same time the Water Authority was faced with the need to replace an outdated sewage works. The Water Authority agreed with the firm that it could have a major share in the new works, asking it to contribute over half of the capital cost (£45,000) and also to pay £4,000 per year in effluent charges, making total annualised costs for the firm of £10,000. The consent conditions set for the firm were a volume of less than 2,000m^3 per day and various physical and chemical maximum concentrations of pollutants, which the management thought could be adhered to.

Prior to the discharge to the new sewage works, the only end-of-line controls the firm employed were neutralisation at a cost of £500 p.a. and the use of a settling tank for precipitation of alumina ferric added to the effluent as a coagulant.

In addition, the firm has taken some in-plant measures to reduce pollution load, principally in the form of eliminating Kier boiling and replacing it with continuous scouring and bleaching. The effect of this was to reduce the liquor ratios, the amount of heat used and the labour input. These changes in processes were, however, taken for economic reasons and not primarily for pollution control. At present the firm has no energy recovery or water reuse, the former because it is technically difficult (although this is now being reexamined in the light of the escalation of energy costs) and the latter because there was no substantial financial consideration for so doing.

The firm has a 45m chimney for its coal fired boilers which are also fitted with a dry grit arrester. The firm spends little on monitoring and research into pollution control, finding water pollution engineers and the Water Authority reliable and helpful in the solution of pollution problems.

We estimate that total costs of pollution control are about 2 percent of total operating costs, which, considering the nature of the effluent, is relatively low. In part this is due to the favourable deal over the new sewage works but the firm has also managed to reduce pollution loads by reducing cloth throughput whilst increasing revenue by upgrading the types of fabrics processed - in other words, by concentrating on quality rather than volume.

This company has made the unusual decision to go "up-market", getting a higher value product and at the same time reducing pollution loads. Yet another solution: none can be the same in this industry. The value of cooperation with the Water Authority again comes through as worthwhile.

Case 21

Whilst we have described this as one case, it is more properly described as a series of five cases carried out in Lancashire. Since the information obtained from each individual firm was limited, it is more appropriate to consider all firms together. The group consists of 3 small firms (with employment of about 50 people or less) and two medium sized firms with the largest employing about 500 people. The output achieved by the firms were roughly in proportion to their sizes in terms of employment, with the small firms producing more than four tonnes per week and the medium-sized plants producing fifty and eighty tonnes per week respectively.

Three plants were handling cotton (or cotton/synthetic mixtures) fabrics and two firms dealt with the dyeing of cotton yarn. The latter were at an advantage over the fabric dyers since they were not involved with any activities such as desizing, scouring, bleaching or mercerising which give rise to fairly high BOD loads. Water consumption varied roughly in line with output; for example, the firm with an output of fifty tonnes per week used about 9,000m³ per week whilst one of the smaller firms used only one twentieth of that amount. Three out of the five firms took their water from boreholes, making up to the total volume with town water or water extracted from a river. The annual cost of water varied considerably, from £400 in one firm to something in excess of £6,000 in another firm, if softening costs were taken into account.

All firms discharged their effluent to the trade sewer although one firm also made a discharge to a river. None of the firms reported any difficulties in achieving, with their existing control plant, the consent conditions stipulated by the Water Authority although one firm which was using sulphur dyes told us that it sometimes had to reschedule its production in order to satisfy the average weekly consent conditions. For the smaller firms, the cost of effluent disposal was of the order of £1,000 p.a. which, whilst small, represented a substantial increase over the prices paid previously and the firms were anticipating costs would rise further in the near future.

In-plant measures to control pollution taken by the smaller firms included chemical substitution (for example, replacing hard detergents with the marginally more expensive, soft biodegradable ones) or changing to high temperature dyeing which had led, in one case at least, to a water saving of the order of 20 percent. None of these firms operated recycling or energy conservation schemes, although one of them had been studying the possibility of the installation of a new German system that would provide water of an acceptable quality for rinsing, it had found that on existing water charges the price would have been prohibitive. In another firm, the use of a heat exchanger would not have been possible, we were told, because of the corrosive nature of the effluent. On the other hand, one of the larger firms reported that it had been able to reduce pollution loads by the more efficient use of chemicals, the installation of a heat exchanger on the dye liquor effluent and the introduction of a counter-current system for water use that had reduced daily water consumption by 225m³ whilst output had increased by almost half over the previous three years.

Four out of the five firms used settling or balancing tanks as end-of-line items of pollution control equipment. In some cases, neutralising chemicals were also added. In one of the larger firms, there was, in addition, the unusual feature of plant to remove solvent from the effluent stream; the solvents were separated and dumped by a subcontractor. In the other larger firm (the one discharging to the river) there was also an activated sludge plant to reduce the BOD load to the required consents; this produced about 4.5m³ of sludge per week which had to be transported for disposal. The annualised costs of operating the end-of-line plant in this firm were about £11,000.

Generally, air pollutants arose only on a small scale; for example stenter fumes would be exhausted to atmosphere and chimneys used for the disposal of sulphide fumes. However, one firm operating coal-fired boilers with a 40m chimney may soon have to instal grit arrestment and another firm experienced considerable difficulty with the disposal of ammonia fumes. In fact, this firm had been prosecuted for this and were under continuing pressure from the Local Authority to reduce them. Even after £12,000 had been spent on fans and scrubbing devices, emissions were still unacceptable and the firm did not know how to cope with them.

We were not able to obtain information from all firms about the proportion of their operating costs accounted for by pollution control but it was one percent in one of the smaller plants and four percent in one of the medium-sized plants. Another of the small plants, reporting total control costs of the order of £15,000 per year said that it would be fairly easy to pass these costs on in the well segmented market it coped with.

<u>It is of course difficult to comment on these firms, but it is interesting to note that, as with a lot of firms, there is reasonable satisfaction with the status quo and no hint of what happens when standards tighten</u>.

Chapter VI

Other Research into Control Costs

Research into the costs of effluent control has not been widespread but in this chapter we review some of the existing British, European, and American work.

THE EAST MIDLANDS STUDY

One study of small firm dyers and finishers in the U.K. (1) attempted to evaluate the impact of pollution controls and effluent charges on prices, employment, market structure and innovation in the industry. It was found that whilst considerable process innovations had occurred among the larger dyers these had not yet permeated down to the smaller firms and water intensive techniques were still dominant. Table 6.1 below summarizes the extent of effluent discharges of this sample of firms.

TABLE 6.1 VOLUME OF LIQUID DISCHARGE BY TEXTILE DYERS AND FINISHERS IN EAST MIDLANDS

Volume of discharge (1,000 gals per week)	Number of firms
Less than 50	6
50 - 249	4
250 - 499	4
500 - 999	4
1,000+	6
	24

Generally it was found that the volumes of effluent were positively associated with size of firm but this relationship was not linear and a decreasing rate of discharges over the total size range was evident, emphasising substantial economies of scale in effluent treatment. The type of effluent varied substantially with the type of fabric and technology used. Some of the newer processes using dispersed dyes which had been developed for dyeing terylene, and transfer printing for synthetics appeared to generate lower volumes of pollution. One problem that was evident was the apparently strong relationship between the quality of the end product and quantity of effluent. In some cases it had been suggested that the quality demands of chain store buyers led to increased dye and rinse water use which ultimately led to an increase in effluent throughput. It was also found that with increases in water charges a remarkable number of firms had made *small* savings in effluent output merely by realigning existing inputs more efficiently.

Fourteen firms out of the total stratified sample treated their effluent in some way. Nine used a settling tank, five used a heat exchanger and only one had a full treatment facility because it was discharging into the river. Although some of the sample claimed that firms operating in other parts of the area had less demanding consent conditions,no evidence was found to suggest the incidence of treatment was in any way related to geographic location. Heat exchangers were sometimes used but these seemed in most cases to be a pollution control device that paid for itself through recycled heat and energy. Most firms had settling tanks which were simply a means for the acid and alkaline wastes to neutralise each other, although sometimes additional chemicals were needed to ensure neutrality of the effluent. Some *in process* changes were found such as Jet dyeing, High temperature dyeing and Continuous dyeing, but these were not widespread and so did not have much impact on the effluent problems of the small firms. It was found that where there was treatment either *in process* or *end of line*,this was something from which economies of scale could usually be derived.

It was found that there was considerable variability in effluent charges and that following reorganisation of local government these may have increased three hundred percent. Even so the charges were still low and it was estimated that they represented only about 5 percent of turnover given the most pessimistic of conditions. However,the authors found that in some cases where air pollution control and hidden/intangible costs of control were estimated,the total cost of abating pollution could approach over ten percent of turnover. The real problems arose when there was a coincidence of pollution control requirements and a low level of corporate profits. Initial cash requirements could mean that only the larger integrated firms could survive in these circumstances.

CONTROL COSTS IN THE U.S.A.

The Environmental Protection Agency have published two studies on control costs in Textile Finishing. The methodology of the first study (2) involved splitting firms up into composite size/financial categories, and then investigating the impact of control legislation on the groups given best practicable technology. To measure the impact of mill closings it was assumed that a mill would close only if the annual cost of treatment exceeded the annual cash flow of the mill. Within this it was further assumed that no mill would close if the estimated incremental operating costs were less than the minimum cash flow. When the operating costs of pollution control were greater than the minimum cash flow but less than the maximum, then it was assumed a proportion of firms would close, the estimated number of plant closings was reckoned as a percentage proportional to the linear relationship between operating costs and cash flow. Thus if the cash flow ranged from $0 to $1,000 and annual operating costs were $250 then 25 percent of all mills were expected to close. For the mills under investigation it was assumed that costs could not be readily passed on to the mills' customers since the mills affected are the smaller independent companies which work in the market for the fill-in needs of the larger customers. It is these latter, the large corporations, who determine what prices to pay, rather in the same way as the big multiple chain stores in the U.K. exercise power over their suppliers. Generally though, costs could only be passed on at all when the market for the final product was strong. Within their sample there appeared very strong evidence that small firms were at a severe disadvantage when dealing with pollution control. The table below translates original data from the study into minimum and maximum control costs as a percentage of selling price for small and medium sized firms in the industry. The study assumed the large integrated companies were either already achieving the proposed effluent guidelines, or if they were not, it was within their financial resources to meet these.

TABLE 6.2 TREATMENT COSTS AS A PROPORTION OF SELLING PRICE IN THE U.S.A. (percent)

	Wool Scouring		Wool Dyeing		Woven Goods Dyeing		Knit Goods Dyeing		Stock and Yarn dyeing	
	s	m	s	m	s	m	s	m	s	m
Minimum treatment cost (if done via municipal sewerage works)	1.5	0.12	1.4	0.7	0.4	0.2	3.0	2.0	3.3	2.0
Maximum treatment cost (best attainable means by the firm on site)	3.5	2.0	3.0	4.0	1.0	0.6	5.0	6.0	6.0	10.0

s = small firm m = medium firm

Source: U.S. Environmental Protection Agency. op cit

The variations are clearly substantial, particularly for those firms who have to do their own processing of waste. The absolute magnitudes, too, are in several instances large, especially for an industry that according to the report is not very profitable. These data led the authors to predict substantial plant closures and unemployment in the industry, particularly in the woollens section.

Further,it is predicted that since many of the smaller plants dominate a local labour market the community effects of these closures may be quite crucial. It was not thought that there would be much impact on international trade.

A second study examined the economic impact on the textile industry of the effluent guidelines for the Best Practicable Technology (B.P.T) the Best Available Technology, (B.A.T.) and the New Source Performance Standards (NSPS), in wool scouring, wool dyeing and finishing, woven fabric dyeing and finishing (cotton and synthetics), knit fabric finishing (cotton and synthetics) and stock and yarn dyeing and finishing. The study was similar to that previously quoted except that in this instance the impact of control costs on prices and employment was more thoroughly investigated, and the methodology of the paradigm predicting impact was amplified. This is particularly interesting and we include a summary below.

(i) <u>Mills closing criterion</u>

The range of cash flows and other financial characteristics of the (composite) mill were estimated and the necessary investment and operating costs were also established; from this the annual operating costs were subtracted from the range of annual positive cash flows. When the annual operating cost did not exceed the minimum cash flow thought to exist for that category it was assumed that no plant closures would ensue from the meeting of the effluent guidelines. The reverse would clearly indicate possible closure.

To make an estimate of the percentage of mills within a category and size classification and in turn an estimate of the number of mills that will close, it was assumed that the cash flow would be distributed linearly throughout the range described. If this assumption is made, then zero or negative cash flow would be proportional to the percentage of mills falling below the point where the annual operating cost would penetrate the cash flow range.

Using the linear distribution the influence that it had upon the results are illustrated in figure 6.1 below. The area defined by "a" in the figure defines the linear distribution where it is equally probable that we will find a cash flow. The area defined by "b" is that of a normal distribution and would express the probability of finding a cash flow within a population obeying this distribution. The difference between these two assumptions can best be seen by comparing "line x" and "line y" which represent the cumulative probability for the linear and normal distributions. In the model it makes no difference whether a normal distribution and/or linear distribution was used at the halfway point. That is to say, both techniques would estimate the same percentage of mill closings.

For cash flows below the mean the linear distribution tended to show a larger number of mill closings than did the normal distribution and, conversely, above the mean the normal distribution tended to increase the percentage of mill closings until finally at the end point where both distributions once again are equal.

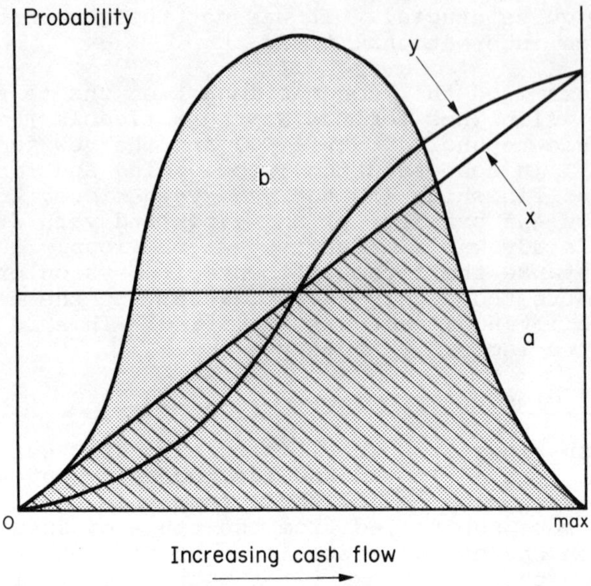

Fig. 6.1

The use of a zero or negative cash flow as the criterion to determine plant closings tended to understate the number of mill closings if higher level criteria were used. Those higher level criteria would include profitability on sales, minimum rate of return on invested assets, internal rate of return, discounted cash flow method, or present value analysis. Each of these forms of analysis would likely carry with it a higher level of cash flow demand to satisfy the criterion and hence would tend to increase the potential number of mill closings.

Because there was an unknown number of mills facing imminent closing decisions, the necessity of meeting the proposed effluent guidelines, and the attendant investments and operating costs would merely hasten the final decision to close the mill rather than actually cause its closing. There would undoubtedly be closings that were ostensibly associated with the mills' inability to meet the investment and operating costs required by the proposed effluent guidelines. Based upon industry sources and interviews undertaken in support of the study, it was estimated that as many as one-half of the mill closings which appeared to occur as a result of the proposed effluent guidelines would simply be accelerated closings. The impacts on employment and production are derived directly from the estimated number of mill closings. The tangential impacts, such as those relative to community pricing, financial and foreign trade were treated for the most part independently.

It was assumed that the larger companies would pass on their treatment costs as a price increase, and that the small and medium size firms would benefit by incrementally greater revenues and thus ameliorate the impact that each would feel as a result of meeting the BAT guidelines. Thus, if competitive conditions prevailing in 1983 permitted large mills to pass on the incremental cost of BAT as a price increase, then the impact on small and medium size mills would be reduced as they too could pass on a proportion of these costs. The proportion that the small and medium size mills are assumed can pass on is equal to the percentage price increase times their annual sales. The resulting increase in revenues may not be able to cover all the incremental BAT costs for small and medium size mills and some could still be impacted and closure may be predicted.

Because the cost of treatment per pound for small and medium size mills may be greater than the cost of treatment by large mills, it was allowed that the small and medium size mills could increase their revenues only by the amount equivalent to the cost of treatment by large mills. Thus if a large mill had a treatment cost of \$0.003/lb, a small mill would have a treatment cost of \$0.004/lb; the latter would be required to absorb \$0.001/lb from its cash flow or other sources.

The analysis was completed by adding the incremental revenue that would be generated by the price increase to the cash flows of each representative size mill. This would result in a new level of cash or funding available in 1983 to defray BAT treatment costs.

(ii) <u>Waste treatment costs</u>

The costs of alternate waste treatment methods to meet the guidelines were calculated on the basis of initial capital investment and annual operating costs for each of the composite mills.

The costs were calculated for best practicable technology waste treatment on the assumption that extended biological treatment would enable a mill to meet the guidelines for BOD, COD, and colour. Best practicable technology (BPT) for the textile industry, as suggested by the EPA, included pretreatment of wastes (screening and 8-hour holdup) and subsequent municipal treatment, or alternatively, self-treatment by preliminary screening, primary settling (wool scouring only), coagulation (carpet mills only), and secondary biological treatment by extended aeration and stabilization ponds.

Pretreatment guidelines for discharge to a municipal treatment system were equivalent to BPT and BAT guidelines with regard to "incompatible" pollutants - COD, total chromium sulphide, phenol, and oil and grease (the latter for wool scouring only). The exception was where the municipal treatment plant was committed in its permit to remove a specified percentage of any incompatible pollutant, in which case the standard for the pollutant was correspondingly reduced. Further, such incompatible pollutants were reduced to acceptable limits by in-plant process changes. Therefore, for the economic analysis, it was assumed that screening and an eight-hour holdup tank represented adequate pretreatment prior to discharge to the municipal waste water treatment plant.

Because of the low annual operating costs associated with the treatment described above, it was further assumed that, on the basis of field visits, such pretreatment was being practised by small and medium size mills that were connected to municipal systems. In addition, it was assumed that new capacity, when built, would most likely take the form of either expansions to existing large mills, or would be new large mills located where self-treatment would likely be the treatment mode. Therefore mills not connected to municipal treatment plants were not considered as a part of the mill closing analysis.

In some unusual cases a mill may be situated in a location which prohibited the use of secondary biological treatment and lagooning, and further may be prevented from discharging to a municipal treatment plant either because the municipal treatment plant did not exist, or it is not large enough. If such conditions existed this would be an unusual condition, the mill would have to decide whether to change its product mix, move its operation, or instal a high-rate activated sludge, two-stage treatment plant.

The space-limited mill condition was not investigated since on investigation it was felt that such situations were so few and had such special conditions that they warranted individual mill-by-mill treatment.

On the basis of this approach the costs and likely impacts of effluent controls were estimated. Some of the costs are included in Table 6.3 below.

By far the most important aspect of this table is the clear advantage in per unit costs of effluent control of the larger mill. In general it was concluded that passing on costs would be difficult for those firms worst affected by the controls (i.e. the smaller independent dyers). Not only were these firms operating in a competitive sellers market, but as in the U.K.,they were often selling to large buyers with considerable market power. The impact of the controls also varied between the different product sectors due to the level of competition operative in the market place. Mainly because of these market power factors it was predicted that the main closures would be in woollen dyeing and stock and yarn dyeing. This would result in a reduction of employment by almost two percent.

TECHNOLOGY AND CONTROL COSTS

Reporting on U.S. and European dyeing technology, a recent article by G. Durig (4) provides some interesting statistics on water consumption and pollution control costs in the textile industry. Arguing that Textiles are an important industry as far as water consumption goes (accounting for approximately 3 percent of industrial water consumption in West Germany (1963) and the U.S.A. (1970)) he shows that the largest part (between 50 and 60 percent) of this is for process water, whilst water for cooling purposes accounts for between 25 and 35 percent and that for steam between 5 and 10 percent. The substantial differences in consumption of the different types of fibre are illustrated in the tables below which are drawn from Durig's work. Not only does the type of fibre have a major part to play in determining the rate of consumption of water, the specific processes used will be of importance *even if the same dyes are used.* For example, in the case of washers for

TABLE 6.3 INVESTMENTS AND OPERATING COSTS TO MEET BPT and BAT GUIDELINES

	Categories											
	Wool Finishing			Woven Finishing			Knit Finishing			Stock & Yarn Dyeing		
Production (1000lb/day)	9.5	33.5	55	9.0	72.5	100	15.0	40	80	11	24	40
Water Use (gal/lb)	13.8	13.8	13.8	18.0	18	18	20	20	20	22	22	22
Treatment plant size (mgd)	0.13	0.46	0.76	0.16	1.3	1.8	0.3	0.8	1.6	0.23	0.53	0.88
BOD Removed (lb/day)	876	3089	5071	236	1900	2620	332	884	1768	520	1040	1732
BPT for mills required to meet normal guidelines ($'000)												
Capital Investment	98	205	278	86	278	342	117	200	304	110	170	227
Annual costs	30	59	79	27	79	96	35	57	86	33	49	65
BPT for mills qualifying as Commission Finishers ($1000)												
Capital Investment	-	-	-	87	284	350	119	205	313	111	172	230
Annual costs	-	-	-	27	79	96	35	57	86	33	49	65
BAT (incremental) ($1000)												
Capital investment	197	349	441	217	570	668	286	452	651	256	372	473
Annual costs	49	89	114	54	152	181	72	118	174	64	95	124

Source: A.D. Little, *op.cit*

TABLE 6.4 CONSUMPTION OF WATER BY FIBRE (l/kg) MATERIAL)

Cotton	50 - 200
Wool	75 - 300
Synthetics	10 - 100

Source: Durig, op.cit

the washing off of reactive dye-prints, the effective liquor ratio for a winch dyer was 1:150, that for a continuous washer 1:50 (a saving of 66 percent) and that for a rotary hydroextraction system 1:6 (a saving of 96 percent over the winch machine).

TABLE 6.5 WATER CONSUMPTION FOR DIRECT DYES (l/kg DYED COTTON)

Method	1	2	3	4
Dyeing	30	10	4	2
Rinsing	60	30	4	3
After-treatment	30	10	4	2
Total	120	50	12	7

1) exhaust method - winch
2) exhaust method - circulating liquor machine
3) exhaust method - jig
4) continuous method - pad-steam

Source: Durig, op.cit

TABLE 6.6 WATER CONSUMPTION FOR VAT DYES (l/kg DYED COTTON)

Method	1	2	3	4
Dyeing	30	10	3	3
Oxidation	30	10	4	4
Soaping	30	10	3	3
Rinsing	100	40	7	7
Total	190	70	17	17

1) exhaust method - winch
2) exhaust method - circulating liquor machine
3) exhaust method - jig
4) exhaust method - pad-jig

Source: Durig, op.cit

TABLE 6.7 WATER CONSUMPTION FOR REACTIVE DYES (l/kg DYED COTTON)

Method	1	2	3	4
Dyeing	20	2	2	2
Rinsing	80	12	14	14
Soaping	20	6	6	6
Total	120	20	22	22

1) exhaust method - winch
2) exhaust method - jig
3) continuous method - pad-steam
4) semi continuous method - cold pad-batch

Source: Durig, op.cit

Durig continues by explaining that ..... "the amounts of dispersed pollutant matter produced in the basic processes of yarn preparation, weaving and knitting (except for synthetic fibres) are largely negligible". The pollution loads vary enormously from process to process - the 5-day BOD loads are highest for wool processing (if the scouring of raw wool is included),next highest for cotton and least for synthetic fibres. In the case of cotton, sizes are the major factor being responsible for almost 75 percent of the pollution load whilst in the case of wool, the natural impurities in the unwashed fibre account for 30 to 50 percent of the pollution load. The potential for the reduction of BOD loads in the case of cotton by changing to a different type of size is illustrated in Table 6.8 which shows the oxygen demand of various sizes.

TABLE 6.8 BIOLOGICAL OXYGEN DEMAND (5-day) OF VARIOUS TYPES OF SIZES

	mg oxygen/kg
Corn starch	810,000
British gum	690,000
Methyl cellulose	1,600
Carboxymethyl cellulose (CMC)	10,000
Poly (vinyl) alcohol	1,600

Source: Durig, op.cit.

The second most important source of pollution after the sizes is the washing and scouring process, followed by the dyes that are used. Here is is argued that dyes "are an insignificant factor compared with other substances entering the environment" - on the grounds that average world-wide dye consumption in the period 1970 to 1972 was of the order of 0.3 million tonnes, compared with over 230 million tonnes of other chemicals (such as nitrogen fertilisers and raw phosphates) entering the environment. However, it is recognised that many dyes contain heavy metals as impurities though "the quantities involved are generally well below the danger levels." The colour problem is such that "man goes to lengths that would seem extreme in the case of colourless chemicals."

TABLE 6.9 COSTS OF CONTROL RELATED TO CONTROL TYPE AND DEGREE OF CONTROL ACHIEVED

Treatment Process	BOD level at end of process	COD level at end of process	Salts pollution load	Proportion of cost of control
	%	%	%	%
None	40	80	85	0
Primary (screening sedimentation neutralisation)	30	65	90	8
Tertiary (physico-chemical treatments)	22	45	88	16
Tertiary (filtration)	22	45	87	24
Secondary (biological oxidation)	0	22	86	50
Tertiary (adsorption)	0	0	85	75
Tertiary (ion exchange osmosis)	0	0	0	100

Source: Durig, op.cit

Some further evidence on the costs of trade effluent disposal may be found in a paper written by G.F. Kitchen for presentation to the Wira Council. (5) In this paper it is suggested that one explanation for the almost tenfold increase in charges over and above those previously calculated by Bradford Corporation was the fact that the oxygen demand of average Yorkshire sewage is only half that of industrial cities making the trade effluent charge twice what it would otherwise have been. The industry responded to the high charges by negotiating concessions from the Yorkshire Water Authority but the author notes that apart from the benefits of these concessions one group of mills was able to save approximately one-fifth of its trade effluent charge by allowing its effluent at least one hour's quiescent settlement thereby reducing the solids content to about the level of sewage. Investigation showed that where new methods had been studied, the cost of pre-treatment was higher than the cost of discharge of trade effluent even after allowing for the recovery of any by-products. This may be largely explained because of economies of scale at the Water Authority plants since the trade effluents are generally of a "similar consistency to domestic sewage."

Finally, our analysis of other research uses some data collected by the Textile Employers Federation, and a compendium of data from a number of plants owned by a U.K. dyer.

THE TRADE ASSOCIATION CASE

This trade association has an intimate knowledge of textile dyeing and finishing in one particular area of the country and was worried about the rapidly escalating level of pollution control costs and effluent charges. For this reason the members had been surveyed and the results are shown in Table 6.10. Since the time of the survey, water charges have been subject to a substantial increase.

TABLE 6.10 WATER COSTS FOR YEAR ENDED 31st MARCH 1975

Cost in Pence/Cubic Metre

	Abstraction & Treatment of Raw Water	Town Water	Effluent	Total	Total Water cost as a proportion of Turnover (percent)
Large Bleach & Dyeworks					
Range	1.7-5.3	5.8-7.9	0.6-7.1	5.7-12.5	2.5-3.6
Average	3.4	7.11	3.8	8.9	3.0
Medium and Small Dyeworks					
Range	1.4-7.4	4.8-9.4	0.6-7.6	4.1-15.0	0.7-3.7
Average	3.5	7.3	3.2	8.7	1.9
Yarn Dyeworks					
Range	1.2-4.3	2.2-10.6	0.8-4.9	3.2-9.6	1.4-1.8
Average	3.0	6.3	2.6	6.3	1.6
Printworks					
Range	0.5-1.2	4.6-15.5	1.7-9.3	3.1-10.4	0.5-2.0
Average	0.8	9.1	4.5	5.7	1.1

10% of works use surface water only
34% of works use town water only
56% of works use both (in most cases town water is less than 10%)
14% of works treat their effluent and discharge to river
24% of works partial treat their effluent and discharge to sewer
62% of works discharge direct to sewer

Source: The British Textile Employers Federation 1976

It appeared that there were two reasons for the rapid escalation in water charges. First there had been a change in the charging formula which now based charges on Chemical Oxygen Demand rather than Biological Oxygen Demand; whilst the water authority was doing all it could to normalise charges under the two systems there were sometimes anomalies and this inevitably led to substantial increases in the price of effluent disposal. Secondly it was felt that the change in control and responsibility for sewage works had led to some of the historical debt of the works being passed to the Water Authorities who now had to reflect this in increased effluent charges.

In addition as the technology of the industry was subject to rapid changes this inevitably led to expertise gaps developing. One effect of this was that for small firms in particular, effluent control through in-plant and in-process changes, was becoming more technologically complex. For the small firms who provided the industry with the essential flexibility of capacity and operation this meant that not only were they subject to rapidly increasing water and effluent disposal costs, but control of their own inputs and outputs was more difficult.

Some of the major in-plant technical changes like high temperature dyeing and solvent cleaning and dyeing, had in some circumstances presented more problems than they had solved since although they had been important in reducing the volume of effluent in some cases they had also changed the quality characteristics of the effluent and this had led to problems for some firms meeting the consents laid down by the Water Authority. In another case a firm had reduced its usage of carriers which benefited the airborne and liquid environment, but found that its effluent was too hot to be discharged to the sewers and was consequently involved in more expense in cooling it.

One other area that was felt to be of major importance was that of colour removal - whilst coloured effluent may be physically harmless it is obviously unpleasant; it creates an amenity pollution problem and may lead to more complaints by members of the public than other discharges. Further, the removal of colour is a technically very difficult procedure.

On the issue of water conservation it was clear that charges will still be a relatively minor component of most firms' costs. Some evidence from other areas had also suggested that substantial water savings had been accomplished through good housekeeping. It was possible that increased water prices might promote this, although product quality parameters would have to be strictly monitored. One factor that could lead to greater economies in the use of water - as well as savings on chemicals, dyes and heat - was the use of a continuous dyeing system but it was felt that this was only feasible in those cases (such as vertically integrated organisations) where it was possible to ensure that supplies of fabric could be standardised. Finally, it was suggested that in spite of the heavily concentrated market structure, small firms could survive by specialising in a segment of the market where only a few firms would compete.

THE COMPOSITE FIRM CASE

This firm owns a large number of plants throughout the country and after our initial approach it decided to send us the results of a survey it had undertaken in 1974 of water sources, consumption and disposal at its plants. This was most useful in isolating plants which had particular problems and some of these we have visited and included in our case studies. We therefore present below the information relating to the other plants that we did not visit and for which we rely on the survey information supplied by the company. We should stress that whilst it is most useful to be able to present a more comprehensive coverage by including these plants, it should be borne in mind that the following 18 are not necessarily representative of the whole industry. All data relate to 1974.

Water usage and sources are shown in Tables 6.11 and 6.12, and it can be seen that there is a fairly even spread of sizes of plant expressed in terms of water usage and the median rate of consumption is 250,000 gallons per week.* Most plants used only water from the town supply and this is also true of the larger plants using more

TABLE 6.11 WATER USAGE (THOUSANDS OF GALLONS PER WEEK)

Consumption	Number of plants
less than 100	4
100 - 250	5
250 - 500	7
500 or more	2
	18

TABLE 6.12 SOURCES OF WATER USED

Sources	Number of plants	Number using more than 250,000 gallons per week
Town only	13	5
River only (including borehole)	2	1
Largely town	-	-
Largely river	3	2
	18	8

* Corresponding to 1,125m³ per week; 1,000 gallons is approximately 4.5m³.

than 250,000 gallons per week (which we shall henceforth call "larger" plants). Only two plants relied entirely on water from a river but three plants used largely river water and these included those using water from boreholes or canals. Three out of the five plants using some river water were larger plants.

TABLE 6.13 WATER TREATMENT BEFORE USE

Treatment	Number	Number using wholly or partly river water
None	13	3
Filtration	3	1
Softening	2	1
Chemical addition	1	-
	19	5

(number sums to more than 18 due to multiple response)

Table 6.13 shows the nature of treatment processes that the water is subjected to before use; in only five firms is the water pre-treated and even in some of the plants using river water, in whole or in part, there is no need for any pre-treatment because, of course, historically raw water quality was a prime consideration in the choice of manufacturing location. In most cases where pre-treatment is needed, it is limited only to filtration or softening; the latter is probably undertaken to improve the quality of the dyed product although it may also reduce some chemical costs. The reason for chemical addition at one of the plants relates to the use of fungicides used to prevent mould growth on delicate machinery.

Tables 6.14 and 6.15 show the means employed by these plants for disposing of their effluent and whether they employ any treatment processes prior to discharge. Most plants discharge direct to the town sewer but two discharged to a river and these were both larger plants. Only two plants had any sort of pretreatment facilities; one of the plants discharging to a river had no form of pretreatment plant.

TABLE 6.14 DISPOSAL METHODS EMPLOYED

Disposal method	Number	Number of larger plants
Town sewer	16	7
River	2	2
	18	9

TABLE 6.15 PRETREATMENT FACILITIES

Method	Number	Number discharging to river
None	16	1
Settlement or acid addition	2	1
	18	2

Tables 6.16, 6.17 and 6.18 show the effluent characteristics in terms of BOD loads, suspended solids and pH; it can be seen that in few plants were very high pollution loads experienced - a factor reflecting the effect of process development designed to minimise pollution. The highest BOD load reported was from a small plant which had a BOD load of 800mg/l. The maximum suspended solids load was 250mg/l in one larger plant. In no plant was a very high or very low pH value recorded and these reflect process design, coupled with chemical recovery in the case of highly alkaline processes such as mercerisation.

TABLE 6.16 EFFLUENT CHARACTERISTICS BOD (mg/l)

Load	Number
Less than 100	2
100 - 200	3
200 - 500	4
500 or more	1
Not available	8
	18

TABLE 6.17 EFFLUENT CHARACTERISTICS SUSPENDED SOLIDS (mg/l)

Load	Number
Less than 100	8
100 - 200	2
200 - 500	2
Not available	6
	18

TABLE 6.18 EFFLUENT CHARACTERISTICS pH

Value	Number
possibly less than 6	3
range 6 - 9	8
possibly more than 9	2
not available	6
	19

(number sums to more than 18 due to one plant having a range 5-10)

In addition to these data on effluent characteristics, the firm also gave us information on items of capital expenditure on effluent treatment plant; this was divided into various categories, the amount spent in the previous 3 years, the amount expected to be spent in the next three years and the amount contributed for local authority treatment plants in the previous 3 years. These data are summarised in Table 6.19 where we also show the sum of these three categories and its distribution for larger plants. In only six plants have there been, or are there to be, no expenditures on treatment plant and most expenditures are concentrated in the larger plants. Of course, it should be remembered that these data relate only to a survey over 3 years and that other major items of capital expenditure had almost certainly been undertaken over a long period before the survey.

TABLE 6.19 CAPITAL COSTS OF POLLUTION CONTROL

Level	Spent in last 3 years	To be spent in next 3 years	Contributions to Local Authorities	Total	Total in larger plants
None	13	10	14	6	2
0-£5,000	3	5	1	6	3
£5-£10,000	1	1	-	-	-
£10-20,000	-	1	1	3	1
£20,000+	1	1	2	3	3
	18	18	18	18	9

It does not seem possible from the data available to relate anticipated capital expenditure to pollution loads although, clearly, there must be some such relationship; the reason for this is that pollution characteristics are not given for the firms with highest anticipated capital expenditure.

TABLE 6.20 WATER PROCUREMENT COSTS

Cost in pence per thousand gallons	Number	Number using wholly or partly river water
Less than 5	4*	4
5 - 10	2	1
10-15	4	0
15-20	2	0
20 or more	6	0
	18	5

* includes 2 plants for which no charge at all was made.

TABLE 6.21 EFFLUENT DISPOSAL COSTS

Cost pence per thousand gallons	Number	Number using trade sewer	Larger plants
None	10	8	7
0-5 *+	4	4	1
5-10 *+	3	3	1
10 or more	2	2	-
	19*	17*	9

* Numbers sum to more than totals due to one firm showing a range of costs between 5.5 and 12.5

\+ 5p/1,000 gallons is just over 1p/m³

Tables 6.20 and 6.21 show the water procurement costs and effluent disposal costs respectively; from the former can be seen the wide range of water charges confronting consumers. Indeed, at two plants water was entirely free of charge; it should be remembered, however, that since the date of this survey the reorganisation of the water supply industry has made some charges increase very rapidly indeed and a more up to date analysis would probably be quite different. Similarly there was a wide range of costs of effluent disposal with over half the plants paying nothing at all, including seven of the larger plants. The two plants discharging into a river also did so without charge. It is interesting that the highest charges were paid by the smaller plants, probably reflecting their lower opportunities for on-site pre-treatment and the differences in effluent characteristics resulting from the precise nature of the fabric types being handled and the production processes involved.

CONCLUSION

The evidence used in this Chapter has been taken from very diverse sources and generalisations are difficult to make. However, the advantages of size, integrated research facilities and the basic heterogeneity of costs are three factors which seem to recur throughout the varied evidence that has been examined.

REFERENCES

1. M. Atkins and J.F. Lowe. Pollution Control Costs in Industry: an Economic Study. Pergamon Press 1977

2. The U.S. Environmental Protection Agency, 'Economic Analysis of Proposed Effluent Guidelines.' E.P.A. 230/1-73-028 1974

3. A.D. Little inc. Economic Analysis of Effluent Guidelines Textiles Industry. National Technical Information Service 1975.

4. G. Durig, "Ecological Aspects of Water Usage in the Textile Industry" in The Review of Progress in Coloration, Vol. 7, 1976

5. G.F. Kitchen, "Trade effluent and water charges: changes arising from the introduction of the 1973 Water Act" Paper presented to W.I.R.A. 1977

Chapter VII
Conclusion

INTRODUCTION

Our case studies analyse the control of pollution in twenty one establishments and cover a wide range of sizes of firm from the smallest (2 employees) to the largest (1,750 employees). We were also able to cover a wide range of process and product types with the exception of bleaching and printing. The sample is,however , very diverse and it would be misleading to make generalisations for the whole industry on the basis of our cases. Instead we prefer, in this chapter, to pinpoint some areas that appear to be important in the context of the major products and processes.

It should also be emphasised that the picture we have drawn of the industry represents only a snapshot at one point of time, textile dyers and finishers have been involved in the control of pollution for many years, their current state has been shaped by the technological alternatives available, and the control stance of public authorities as well as their own social responsibility.

Prior to 1974 the control of water pollution from this industry was split between two bodies; the Local Authorities who had responsibility for discharges through the sewerage and the River Authorities who had responsibility for discharges direct to rivers. Since then, however, ten Water Authorities have been formed and assumed complete control of all discharges. All discharges to rivers are subject to consents agreed between the Water Authority and the firm, generally these will involve the firm in bringing its effluent quality to a minimum level. There are no charges made for effluent discharged to rivers. For discharges to the sewerage consents are also operative but charges are made on the basis of the cost of effluent conveyance and treatment. There are no widely published details of the nature of the consents *which are operative* but they will cover parameters such as rates of flow, B.O.D./C.O.D. load, metal content and effluent temperature. Some authorities may also lay down requirements concerning effluent colour although there seems to be no universally adopted standard for this. Indeed there is something of a lack of uniformity in standards and charges between authorities, although since the 1973 Water Act Section 30(5) previously referred to variations will be reduced although will still persist where there are different costs of treatment, environmental damage, sewage works capacity and other socio/economic considerations. Inevitably some firms will have to pay more for effluent treatment.

THE DETERMINANTS OF POLLUTION OUTPUTS

The single most important determinant of the nature of the firm's pollution problem is the precise nature of the type of fabric or fibre that is being dyed; the dyeing of synthetic or man-made fibres generally leads to fewer pollution problems than the dyeing of natural fibres although it does sometimes bring about difficulties not experienced in the dyeing of natural fibres. Within the sector involved with dyeing natural fibres, two principal sources of pollution can be identified. The first is scouring raw wool which gives rise to very large amounts of grease and consequently very high BOD loads and suspended solids ratings, but where the credit from recovered lanolin may, in fact, pay the greater proportion of the cost of control. The second is the de-sizing operation carried out by dyers of cotton and which can give rise to large quantities of starch in the effluent, although the volumes involved in this operation are small in comparison to the firm's total usage of water. Scouring is again the most important source of BOD load in this sector. In the dyeing and finishing of synthetic fibres, the rate of consumption of water can be reduced by a switch to high temperature dyeing and whilst this may bring about some quality benefits as well, in certain circumstances the temperature of the effluent may be adversely affected. The precise characteristics of the firm's pollution problem may also depend on whether it is dyeing yarn or cloth; in some cases the former may give rise to greater pollution loads because of the discharge of large amounts of fibres into the effluent streams resulting in high suspended solids ratings. A further technical factor determining the firm's pollution load will be the dyes it chooses to use within the constraints of the production processes which, of course, also depend upon fibre types.

We are also able to identify various structural and economic factors which may influence the firm's pollution load. Firms which dye on commission tend to be unable to reap the benefits of production planning and consequent reduction in pollution loads that in-house dyers are able to achieve. These economies arise in two ways; in-house dyers tend first to have longer production runs and secondly a more homogeneous range of inputs and outputs. Both these factors enable the in-house dyer to indulge in better housekeeping than the commission dyer which could, on the basis of our evidence suggest a saving of 20 percent of water and dyestuff consumption. In fact, this situation has received formal recognition in the U.S.A. where less stringent effluent guidelines apply to commission finishers: "to allow for a greater variability of wastes since they are expected to handle a wider variety of fibres, fabrics and dyes and to operate on much shorter production runs".(1) In spite of these less stringent guidelines, anticipated effluent treatment costs are as high as those for other groups. Further, the larger integrated firms with their greater access to financial resources and to resources of technical and research expertise will find it easier to install the least pollution intensive equipment like the rotary hydroextraction system that can result in water savings of 96 percent. (2) The evidence of the cases also indicates that water procurement costs may act as incentives to reducing water consumption and hence pollution outputs. This is likely to be an important factor since there were many firms in our sample which abstracted water from rivers or boreholes at zero or only nominal cost.

AN OVERVIEW OF THE CASES

An overall analysis of our cases reveals several factors which it is felt are significant in an analysis of the way in which this industry has dealt with its pollution problems. Firstly it was found that there was considerable variability in effluent disposal costs even when account was taken of different volumes of discharge in large and small firms. These costs appear to be substantial but small in relation to the firm's total cost of production and unlikely to exceed 5 percent of that. Nevertheless, many firms experienced rapid increases in these charges (up to 900 percent over 1975/76) at a time when profitability in the industry was exceptionally low.

For those firms not connected to local authority sewage works and with their own treatment plant, costs of disposal are higher although the recent inflation in effluent charges has meant that at least one of the large integrated groups has considered building its own treatment facilities even when a local authority sewage works can be used. Most firms discharging to the sewerage had some facilities for partial treatment of their effluent but had been unable to assess any saving in trade effluent charges that arose from this. The installation of partial treatment is usually a precondition to the granting of a consent to use the sewer rather than a way of saving on effluent charges.

In general, pollution loads may be reduced by a factor of approximately one half by relatively simple means such as screening or equalisation but higher standards of control will inevitably require a more sophisticated chemical engineering approach with much higher costs. The greatest proportion of those firms which at present did not meet consent conditions dye cotton and this may be attributed to the use of low-cost dyestuffs containing sulphur. For example, we have been told of one such case where the mean values of a number of samples of the effluent from the firm revealed that consent conditions for suspended solids were being exceeded by a factor of 100 percent, for BOD by a factor of 5000 percent, for sulphates by a factor of 900 percent and for temperature by 16 percent. It had been estimated that control costs would be likely to exceed £100,000 p.a. Imposing strict standards on *all* firms could therefore create more severe problems in regions where cotton dyeing and finishing is concentrated which happen to be ones where there is already a high level of unemployment.

Effluents from textile mills are generally non-toxic and control may be easily operated through the parameters of BOD, suspended solids, Ph and temperature. However, in the case of the cotton industry where sulphides, sulphates or even chrome may be present in the effluent, further controls may be imposed.

Unlike other industries there is little scope in textile dyeing and finishing for the generation of recovery credits through pollution control and in most cases the full cost of pollution control must be paid for; the principal exception to this was case 1 where lanolin could be recovered from the wool scouring process. In many instances the firm may be able to recover heat from its effluent and many firms had installed or were considering such equipment. Obviously further increases in the cost of energy are likely to stimulate moves in this direction and it has been estimated (3) that for a cheese-dyeing machine of about 3000 litres capacity heat saving may amount to about three-quarters of that needed to heat the dye liquor to 95°C.

It is a feature of this industry that pollution control in the firm has taken the form of in-plant measures more often than end-of-line controls. For example, using a continuous dyeing machine rather than a direct jig dyeing machine may result in water savings of the order of 70 percent, using a jet dyeing machine would result in savings of between 30 and 50 percent by comparison with a winch dyeing machine or by changing from corn starch size to CMC (carboxy methyl cellulose) 5-day BOD loads may be reduced from 810,000 mgO_2/kg to 10,000 mgO_2/kg.* (4) The use of sophisticated (and land consuming) end-of-line controls is at present limited almost entirely to those firms discharging direct to water courses although we anticipate that if the potential economies of total water recycling improve further, this situation may well change in the future.

The pollution problems of textile dyers and finishers are not restricted to liquid effluent problems. In a number of cases air pollution may arise from the use of stentering machinery and this could become a major source of expenditure soon. Fumes may be ducted to atmosphere and discharged at high level or it is possible to arrange for them to be burnt. The capital cost of this would be large although it could result in savings on energy costs for stenters.

ECONOMIC CONSIDERATIONS

An objective of this book has been to emphasise that the economics of pollution control in the industry must be evaluated on a case by case basis if a sensible overall view is to be formed. However there are some important issues that arose in the cases that can be related to the basic economic factors discussed in the first chapter.

Where a firm does not have its own treatment facilities it is only subject to the charges levied on it by the water authority; these costs vary with the rate of output and should in theory be neutral between firms of different sizes. This is less true where firms have their own pretreatment facilities since costs of construction and manufacture are subject to economies of scale. Table 7.1 below lists the capital and operating cost components of treatment equipment where this has been installed.

---

* The change from corn starch to C.M.C. may reduce B.O.D. but increase C.O.D., when charges are based on the latter this may in fact lead to an increase in the effluent bill.

TABLE 7.1 COST BREAKDOWN OF END-OF-LINE TREATMENT

Case No.	Capital costs £	Operating costs (per annum) £
1	250	100
14	50	20
16	17.5	5
others not in case material	250	18
	250	28
	20	28
	50	5.5

The breakdown of these costs is largely in line with U.S. experience and the substantial capital component clearly represents a source of potential entry barriers in the industry if own treatment becomes more prevalent.

As indicated previously, unsophisticated control measures are relatively inexpensive, however as the degree of control required increases and more complex technologies are necessitated costs of control increase rapidly. For instance Durig (5) shows that the introduction of biological oxidation may double control costs and if this were required together with adsorption, ion exchange and osmosis, control costs may be quadrupled for a marginal benefit of reduced BOD load (25 percent to zero) and COD load (45 percent to zero).

The cost breakdown of in-plant measures is likely to be even more dominated by the capital element, however as it was difficult to separate the pollution control benefits of in-plant measures from the production benefits, we have not been able to make many meaningful estimates of this.

Ultimately the real costs of control are impacted by a large variety of non quantifiable factors which are nonetheless important. In particular we found that age of plant, site area, specific production processes, plant location, nature of the water source and structural organisation of the enterprise all affected the cost structure of effluent control. There were some costs that resulted from effluent control necessitating certain profitable or potentially profitable activities to be foregone. Treatment plants that took up valuable production space were one important example that we found under certain assumptions could substantially increase the true costs of control; whilst in one instance a firm had decided not to pursue the production of a new fire resistant material because of the effluent problems it would generate. These 'opportunity costs' are likely to be substantial in some instances but their measurement is difficult. It is therefore not surprising that the Environmental Protection Agency acknowledges these costs in their impact studies but does not quantify them.

Economies of scale in pollution control arose essentially from two sources. Firstly there were the plant economies that were important in the operation of particular pieces of equipment *once these had been installed*. Cases 1, 16 and 20 were the only ones operating full treatment and it is important to note that these were among the largest firms in our sample. Generally the closer the firm wanted to move towards full treatment the larger the throughputs had to be in order to justify a full treatment system. Our research did not include detailed discussions with pollution control manufacturers and it is not meaningful to use the case data on a cross sectional basis to estimate scale economies of pollution control. However an inspection of cost data from the Environmental Protection Agency for multi media filtration, equalisation, and aerated stabilisation shows clearly evidence of plant economies operating in environmental control. Table 7.2 below shows that in general increasing returns to scale operate over the full range of operational outputs. Whether or not the impact of these scale economies would increase the minimum efficient size of firm will, of course, depend on whether the optimum level of pollution control occurs at higher or lower output rates than the optimum level of production. The case studies certainly support the qualitative impression that pollution control does lead to an increase in the minimum efficient size of plant and quantitative support for this can be found in U.S. research. (6)

TABLE 7.2 EFFLUENT CONTROL COSTS AND SCALE OF OPERATION

	Flow rate of effluent (millions of gallons per day)		
Total control costs ($)	0.1	1.0	10.0
Equalisation			
Initial capital cost	23,732	75,100	248,107
Annual variable cost	203	256	542
Multi media filtration			
Initial capital cost	36,000	180,000	800,000
Annual variable cost	100	540	1,400
Aerated stabilisation			
Initial capital cost	16,000	35,000	150,000
Annual variable cost	1,008	3,208	9,856

Source: A.D. Little op.cit

The other major source of scale benefits existed in the context of centralised research and development facilities. Whilst the inventiveness of the small entrepreneurial unit had often been a noticeable feature of the growth of capitalist economies we found that the small firm in this industry was at a distinct disadvantage when attempting to comply with changes in environmental regulations. Sometimes this was due to an inability to communicate effectively with the relevant control bodies, but it also seemed evident that pollution control represented a new technology that had to be implemented irrespective of commercial considerations and in this area of applied research and development the small firm was at a distinct disadvantage. On the basis of our case study evidence we estimate that the recourse to centralised research facilities was one of the most important aspects of control cost reduction. It was also an important determinant for generating the greatest benefits from recycling.

The major resource savings in the industry are in heat, water and chemicals. Recently the balance of these has been altered with the introduction of various process changes (which as a side effect have also impacted pollution loads) such as high temperature and solvent dyeing which have reduced the need for large quantities of water but which have raised effluent temperature and may introduce new insoluble liquid contaminants. However, overall the majority of dyeing processes were still the traditional water and heat intensive methods and it is the recycling and reuse of these that can be analysed.

Only 8 firms failed to be involved with any form of recycling whilst only 6 had full recycling programmes for both liquid inputs and heat. Thus recycling was a fairly common characteristic of the firms' operations but was not something many firms were *heavily* involved with.

A priori it would be reasonable to expect that the size of establishment would be an important determinant of recycling because typically for recycling to be worthwhile there are usually production economies of scale involved and a larger firm might also be expected to be more capable in the assimulation of the necessary technology. Using Rees'(7) minimum efficient size of plant estimate of 100 employees we can relate the size of firm above and below this to the extent of recycling and use the chi squared distribution to test for statistical association (8).

Our results showed that the size of firm was relatively unimportant as a determinant of whether the firm will recycle water and heat. Some of the largest firms were not involved in recycling, although if the size cut-off point is increased to 400 employees then almost 70 percent of recycling possibilities are taken up by these *very* large establishments. However, all of these firms are also parts of larger groupings, and to investigate the effect of organisation type in this integrated industry we looked at the extent of recycling between integrated and independent firms.

The result of a χ^2 test here showed that we can be almost 95 percent sure that the extent of vertical integration is positively associated with recycling of either water or heat.

Most of the heat and water recycling systems were of a relatively simple nature, the problem for firms was not of affording them (although the more sophisticated water-reuse systems were expensive) but of understanding and being able to use them. Mis-specification and misunderstanding of pollution control equipment was common; one small firm we talked to had spent £15,000 on a treatment facility that had never been used, due to mis-specification of the effluent parameters.

Pressure for the firm to recycle and conserve resources also arises on the demand side from an inflation in raw material prices, effluent charges, or control agency pressure concerning standards. We found that the average effluent charge for firms recycling was 34.1p per thousand gallons whereas those not recycling only had a mean of 8.8p per thousand gallons.

Recycling of certain inputs, materials, etc. is increasing in importance but its extent is limited by financial feasibility, the extent of technological know-how existing within the enterprise and to a lesser extent, the size of the firm and possible scale economies. At present much recycling in this industry has necessitated fairly cheap inputs or merely minor process adjustments (which have nevertheless involved a fair understanding of the technology involved). However, the more sophisticated water and heat recycling equipment seem likely to involve large capital expenditures and are only likely to be adopted if effluent charges, water and heat costs increase very much more than at present. In those situations it is likely that many smaller firms would be in a disadvantageous position.

The evidence just summarised on cost structures, capital requirements, barriers to entry and economies of scale suggest that an increase in concentration and a reduction in the number of enterprises is likely in an industry that is already quite heavily concentrated. Thus in the short term there are likely to be firms leaving the industry but in the longer term if the industry used its market power to pass on the increases in pollution control costs the situation would soon stabilise as long as it can be assumed that foreign competition is limited in some way.

However it appears that for the large majority of firms in this industry, particularly those using local authority sewers, pollution problems are not large and effluent disposal costs are not high although as a proportion of profit they may be substantial and have been subject to rapid and large increases. A minority of plants, largely those discharging directly to rivers, do have substantial disposal problems and high disposal costs but to an extent these firms are cushioned from the water authorities rising charges. Larger firms may well be in a much better position to withstand higher disposal costs and pollution standards than small firms.

REFERENCES

1. M. Atkins and J.F. Lowe. Pollution Control Costs in Industry: an Economic Study. Pergamon Press 1977

2. The U.S. Environmental Protection Agency, 'Economic Analysis of Proposed Effluent Guidelines.' E.P.A. 230/1-73-028 1974

3. A.D. Little Inc. Economic Analysis of Effluent Guidelines: Textiles Industry. National Technical Information Services 1975

4. G. Durig, "Ecological Aspects of Water Usage in the Textile Industry" in The Review of Progress in Coloration, Vol.7, 1976

5. G.F. Kitchen, "Trade effluent and water charges: changes arising from the introduction of the 1973 Water Act" Paper presented to W.I.R.A. 1977

REFERENCES

1. Arthur D Little, Inc. "Economic Analysis of Effluent Guidelines: Textiles Industry" Environment Protection Agency 1975

2. G. Durig, "Ecological Aspects of Water Usage in the Textile Industry" in Review of Progress in Coloration, Vol.7, 1976

3. G. Durig, ibid

4. G. Durig, ibid